Lecture Notes in Control and Information Sciences

Edited by A.V. Balakrishnan and M. Thoma

27

D. H. Jacobson · D. H. Martin
M. Pachter · T. Geveci

Extensions of Linear-Quadratic Control Theory

Springer-Verlag
Berlin Heidelberg New York 1980

Authors
D. H. Jacobson · D. H. Martin · M. Pachter · T. Geveci
National Research Institute for Mathematical Sciences,
Council for Scientific and Industrial Research,
P.O. Box 395,
Pretoria 0001
Republic of South Africa

Some of the material included in these Lecture Notes stems from research work at NRIMS which was partially supported by a grant from Control Data.

ISBN 3-540-10069-5 Springer-Verlag Berlin Heidelberg New York
ISBN 0-387-10069-5 Springer-Verlag New York Heidelberg Berlin

Printed in Germany

Printing and binding: Beltz Offsetdruck, Hemsbach/Bergstr.
2061/3020-543210

PREFACE

These Lecture Notes were prepared for presentation as a Summer Seminar Series on Extensions of Linear-Quadratic Control Theory, at the National Research Institute for Mathematical Sciences (NRIMS) of the Council for Scientific and Industrial Research, Pretoria, during 4-8 February 1980. However it is felt that these notes will be of interest and value to a wider audience, both for the concise survey of much of the theory of multi-variable linear-quadratic optimal control and estimation, and for details of various extensions to the theory developed over the past five years at NRIMS.

While each lecture naturally contains some basic subject matter, the first six lectures and the last two are in the nature of comprehensive surveys, covering material that is somewhat scattered in textbooks, re=search monographs and research articles. On the other hand, Lectures 7 through 13 present the results of more recent research done in the Mathe=matics Division of NRIMS. The style presupposes some familiarity with real vectors and matrices, with systems of ordinary differential equa=tions, and in Lectures 12 and 15, with a little basic Hilbert space theory. However the flavour is decidedly applied mathematical, and these notes should be of use principally to those concerned with process control in industry, dynamic modelling and decision theory in economics and management science, and to members of departments of applied mathe=matics and engineering at universities and technical colleges.

As editor of this volume, it is a pleasure to thank Mrs. M. Russouw of NRIMS for the very competent way she has transformed our manuscripts into beautifully typed pages.

D.H. Martin (Editor)
Pretoria 1980.01.10

CONTENTS

1. REVIEW OF LINEAR DYNAMIC SYSTEMS

1.1 CONTINUOUS-TIME DYNAMIC MODELS

This Summer Seminar Series concentrates on the control of linear dyna= mic systems, and consequently we shall be concerned primarily with the properties of these systems. However, nonlinear differential equations (actually matrix Riccati equations) arise in the solution of linear-quadratic control problems, so it is useful to be aware of certain phenomena, not present in the linear case, which arise in the solution of nonlinear differential equations.

A more basic and more generally important motivation for including non= linear ordinary differential equations is that they often arise as models of real world systems. Fortunately, linearized versions of non= linear models at an operating point, or along an operating path, are often adequate to determine controllers for the real world plant, valid for initial conditions which are 'near' to the nominal operating point/ path at which the linearized model is determined.

In view of the above points we proceed now to a brief review of non= linear ordinary differential equations.

1.1.1 Continuous-time nonlinear dynamic systems

A rather broad class of continuous-time mathematical models of real world dynamic systems is that which is characterized by the following simultaneous nonlinear ordinary differential and algebraic equations defined for $t \in [0,\infty)$:

$$\dot{x}_i(t) = f_i(t,x_1(t),x_2(t),\ldots,x_n(t),u_1(t),u_2(t),\ldots,u_r(t)), \quad i=1,\ldots,n \tag{1.1.1}$$

$$y_j(t) = g_j(t,x_1(t),x_2(t),\dots,x_n(t)), \quad j=1,\dots,m \qquad (1.1.2)$$

where the initial conditions $x_i(0)$ are specified as

$$x_i(0) = x_{i0} \ , \quad i=1,\dots,n. \qquad (1.1.3)$$

The n differential equations (1.1.1) model the dynamic behaviour of the system as it evolves in time as a consequence of both the effects of the initial conditions (1.1.3) and the input or control functions $u_1(\cdot),\dots,u_r(\cdot)$, while the m algebraic equations (1.1.2) produce the m system outputs $y_1(t),\dots,y_m(t)$ via appropriate nonlinear functions of the x-variables. Clearly, knowledge of $x_1(t),\dots,x_n(t)$ and $u_1(\tau),\dots,u_r(\tau)$, $\tau \geqslant t$ is all that is necessary to determine $y_1(\tau),\dots,y_m(\tau)$, $\tau \geqslant t$; consequently $x_1(t),\dots,x_n(t)$ is referred to as the *system state* at time t.

In the real world one knows (measures) only the input function $u(\cdot)$ and the output function $y(\cdot)$ of a system. The model (1.1.1-3) is therefore referred to as a *realization* of the input-output relationship observed in the real world. The state vector x (which is not unique) and the form of the 'state machine' (1.1.1) are normally suggested to the modeller by the physics, biology, economics, etc. of the real world process. Note that in the real world one does not normally observe $x(t)$ directly, but only implicitly via the system outputs $y(t)$.

For ease of handling we shall write the above equations in vector form as follows:

$$\dot{x}(t) = f(t,x(t),u(t)), \quad x(0) = x_o \qquad (1.1.4)$$

$$y(t) = g(t,x(t)) \qquad (1.1.5)$$

where $x(t) \in \mathbb{R}^n$, $u(t) \in \mathbb{R}^r$, $y(t) \in \mathbb{R}^m$. Clearly,

$$x(t) = \begin{pmatrix} x_1(t) \\ \vdots \\ x_n(t) \end{pmatrix},$$

and similarly for the other vectors and vector functions in (1.1.4) and (1.1.5).

As stated, (1.1.4) and (1.1.5) are purely formal equations for which a solution need not exist. We therefore introduce the following assumptions.

ASSUMPTION 1.1.1 The vector functions $f(t,x,u)$ *and* $g(t,x)$ *are continuous in* t *and once continuously differentiable in* x *and* u, *for fixed* t, $t \in [0,\infty)$.

ASSUMPTION 1.1.2 The control functions $u(\cdot)$ *are drawn from the class of piecewise continuous* m*-vector functions defined on* $[0,\infty)$.
Under these assumptions we have the following theorem.
THEOREM 1.1.3 [1] *The differential equation* (1.1.4) *has a unique solution* $x(t)$ *defined for* $t \in [0,\bar{t}]$, $\bar{t}$ *sufficiently small.*

Note that nonlinear differential equations can cease to have a solution after the lapse of a certain interval of time, as the following example illustrates.

EXAMPLE 1.1.4 The differential equation

$$\dot{x}(t) = x^2(t), \quad x(0) = \varepsilon > 0 \qquad (1.1.6)$$

satisfies Assumptions 1.1.1 and 1.1.2 and has a unique solution

$$x(t) = \frac{\varepsilon}{1-\varepsilon t}, \qquad (1.1.7)$$

in accordance with Theorem 1.1.3. However, this solution ceases to exist at $t = \frac{1}{\varepsilon}$.

1.1.2 Linearization

It is not easy to calculate control functions $u(\cdot)$ which force general nonlinear models of the form (1.1.4-5) to behave as desired. Usually one has to be content with the statement of certain necessary conditions which must be satisfied by such control functions, and which have then to be solved numerically by iterative means.

The situation is much brighter if the system is linear, and fortunately one is often interested in the behaviour or control of (1.1.4-5) only in the neighbourhood of a desired operating path $y_e(t), x_e(t), u_e(t)$, $t \in [0,\infty)$. By definition of an operating path, we have

$$\dot{x}_e(t) = f(t, x_e(t), u_e(t))$$

$$y_e(t) = g(t, x_e(t)),$$

and defining

$$\xi(t) = x(t) - x_e(t)$$

$$\eta(t) = u(t) - u_e(t)$$

$$\mu(t) = y(t) - y_e(t)$$

we have

$$\dot{\xi}(t) = f(t, \xi(t)+x_e(t), \eta(t)+u_e(t)) - f(t, x_e(t), u_e(t)) \tag{1.1.8}$$

$$\xi(0) \triangleq \xi_o = x(0) - x_e(0)$$

$$\mu(t) = g(t, \xi(t)+x_e(t)) - g(t, x_e(t)). \tag{1.1.9}$$

For sufficiently small ξ_0, $\xi(t)$, $\eta(t)$, it follows, upon expanding the right-hand sides of (1.1.8), (1.1.9) about $x_e(t)$, $u_e(t)$ - which is permissible by Assumption 1.1.1 - that $\xi(t)$ and $\eta(t)$ are *approximately* given by

$$\dot{\xi}(t) = \frac{\partial f}{\partial x}(t,x_e(t),u_e(t))\xi(t) + \frac{\partial f}{\partial u}(t,x_e(t),u_e(t))\eta(t) \tag{1.1.10}$$

$$\xi(0) = \xi_0$$

$$\mu(t) = \frac{\partial g}{\partial x}(t,x_e(t))\xi(t). \tag{1.1.11}$$

Thus, provided deviations $\xi(t)$, $\eta(t)$ from $x(t)$, $u(t)$ are not too large, it may be hoped that the linear model (1.1.10-11) will adequately de= scribe the behaviour of (1.1.4-5). This hope turns out to be justified under certain important circumstances, some of which we illustrate in Section 1.4.

1.2 CONTINUOUS AND DISCRETE-TIME LINEAR DYNAMIC SYSTEMS

We rewrite (1.1.10) and (1.1.11) in the more compact form

$$\dot{x}(t) = A(t)x(t) + B(t)u(t), \quad x(0) = x_0 \tag{1.2.1}$$

$$y(t) = C(t)x(t) \tag{1.2.2}$$

where $A(t) \in \mathbb{R}^{n \times n}$, $B(t) \in \mathbb{R}^{n \times r}$, $C(t) \in \mathbb{R}^{m \times n}$ are piecewise continuous matrix functions of time.

It turns out that as $A(\cdot)$ and $B(\cdot)$ are piecewise continuous functions of time, (1.2.1) has a unique solution defined for all $t \in [0,\infty)$. We state this important result as a theorem.

THEOREM 1.2.1 [2] *If the matrices* $A(\cdot)$ *and* $B(\cdot)$ *have elements which are piecewise continuous functions of time on* $[0,\infty)$, *then* (1.2.1) *has a unique solution defined on* $[0,\infty)$.

1.2.1 Solution of continuous-time linear dynamic systems

An explicit solution for the inhomogeneous equation (1.2.1) can be written in terms of the solution to the associated homogeneous matrix differential equation

$$\frac{d}{dt}\Phi(t,\tau) = A(t)\Phi(t,\tau), \quad \Phi(\tau,\tau) = I. \qquad (1.2.3)$$

The matrix $\Phi(t,\tau)$ is known as the transition matrix for (1.2.1).

By direct differentiation it is easy to confirm that the solution of (1.2.1) is

$$x(t) = \Phi(t,0)x_0 + \int_0^t \Phi(t,\tau)B(\tau)u(\tau)d\tau. \qquad (1.2.4)$$

EXAMPLE 1.2.2 Newton's law of motion for a unit mass is

$$\ddot{x}(t) = u(t). \qquad (1.2.5)$$

Rewritten in our standard format this is

$$\dot{x}_1(t) = x_2(t)$$
$$\dot{x}_2(t) = u(t)$$

or

$$\begin{pmatrix}\dot{x}_1(t)\\ \dot{x}_2(t)\end{pmatrix} = \begin{pmatrix}0 & 1\\ 0 & 0\end{pmatrix}\begin{pmatrix}x_1(t)\\ x_2(t)\end{pmatrix} + \begin{pmatrix}0\\ 1\end{pmatrix}u(t). \qquad (1.2.6)$$

Using (1.2.4), and noting from (1.2.3) that in this case $\Phi(t,\tau) = \begin{pmatrix}1 & t-\tau\\ 0 & 1\end{pmatrix}$, we have Newton's law in integral form as

$$x(t) = x(0) + \dot{x}(0)t + \int_0^t (t-\tau)u(\tau)d\tau. \qquad (1.2.7)$$

The most notable properties of the transition matrix $\Phi(t,\tau)$ are the following.

THEOREM 1.2.3

(i) $\Phi(t,\tau)$ *is invertible,* $t, \tau \in [0,\infty)$;

(ii) $\Phi(t,\tau) = \Phi(t,\bar{t})\Phi(\bar{t},\tau)$;

(iii) $\Phi(t,\tau) = \Phi^{-1}(\tau,t)$;

(iv) *if* A *is a constant matrix, then* $\Phi(t,\tau) = e^{A(t-\tau)} = \sum_{k=0}^{\infty} \frac{1}{k!} A^k(t-\tau)^k$.

PROOF Consider

$$\frac{d}{dt}\psi(t,\tau) = -\psi(t,\tau)A(t), \quad \psi(\tau,\tau) = I.$$

Now

$$\frac{d}{dt}[\Phi(t,\tau)\psi(t,\tau)] = A(t)\Phi(t,\tau)\psi(t,\tau) - \Phi(t,\tau)\psi(t,\tau)A(t) \quad (1.2.8)$$

with

$$\Phi(\tau,\tau)\psi(\tau,\tau) = I.$$

Equation (1.2.8) is a linear differential equation in

$$Z(t,\tau) = \Phi(t,\tau)\psi(t,\tau)$$

whose coefficients are continuous functions of time. Consequently the equation has the unique solution $Z(t,\tau) = I$.

We have also that

$$\frac{d}{dt}[\psi(t,\tau)\Phi(t,\tau)] = 0,$$

so that $\psi(t,\tau)\Phi(t,\tau)$ is also the unit matrix I. Accordingly $\psi(t,\tau)$ is the inverse of $\Phi(t,\tau)$, and (i) has been proved by construction.

Property (iii) follows from (ii) upon setting $\tau = t$ and $\bar{t} = \tau$, while the proof of property (ii) is left as an exercise. For (iv), see [2].

1.2.2 Discrete-time linear dynamic systems

In the computer implementation of control theory it becomes necessary to discretize at one or another stage. One possibility is to discretize (1.2.1), (1.2.4) to obtain a discrete-time dynamic system. This is accomplished by holding u(t) constant over an interval of time of length T, i.e.

$$u(t) = u_k, \quad kT \leq t < (k+1)T. \tag{1.2.9}$$

Using this controller in (1.2.4), and also making use of property (ii) of Theorem 1.2.3, we have

$$x[(k+1)T] = \Phi[(k+1)T, kT]\,x(kT) + \int_{kT}^{(k+1)T} \Phi[(k+1)T, \tau]\,B(\tau)\,d\tau\, u_k \tag{1.2.10}$$

or, in an obviously simpler notation,

$$x_{k+1} = A_k x_k + B_k u_k, \quad x_o \text{ given.} \tag{1.2.11}$$

Accordingly, the measurement equation is

$$y_k = C_k x_k. \tag{1.2.12}$$

1.3 PROPERTIES OF CONTINUOUS-TIME LINEAR DYNAMIC SYSTEMS

In this section we review conditions for the controllability, observability and stability of (1.2.1) and (1.2.2).

1.3.1 Controllability

There are many slightly different notions of controllability; the one we introduce here concerns the steering of the system state to the origin in a specified time.

DEFINITION 1.3.1 System (1.2.1) *is completely controllable over the time interval* [0,T] *if any initial condition* x_0 *at time* $t = 0$ *can be steered to the origin of the state space, viz.* $x = 0$, *at time* $T < \infty$ *by a continuous control function* $u(\cdot)$.

THEOREM 1.3.2 A *necessary and sufficient condition for complete controllability of* (1.2.1) *over* [0,T] *is that the matrix*

$$W(0,T) \triangleq \int_0^T \Phi(T,\tau)B(\tau)B'(\tau)\Phi'(T,\tau)d\tau \text{ is positive definite.} \tag{1.3.1}$$

PROOF If W(0,T) is positive definite, it is invertible. Consequently we set

$$u^*(t) = -B'(t)\Phi'(T,t)\,W^{-1}(0,T)\Phi(T,0)x_0, \quad t \in [0,T] \tag{1.3.2}$$

and substitute this into (1.2.4). We then find that

$$x(T) = \Phi(T,0)x_0 - \int_0^T \Phi(T,\tau)B(\tau)B'(\tau)\Phi'(T,\tau)d\tau W^{-1}(0,T)\Phi(T,0)x_0 = 0.$$

This proves the sufficiency part by actually constructing a control (1.3.2) which does the job of steering the system to the origin.

Suppose now that the system is completely controllable but that W(0,T) is not positive definite. Then there is a non-zero vector $\bar{x}$ such that

$$\bar{x}'W(0,T)\bar{x} = 0, \tag{1.3.3}$$

which implies that

$$B'(t)\Phi'(T,t)\bar{x} = 0, \quad t \in [0,T]. \tag{1.3.4}$$

Now since the system is completely controllable, there is a u(•) such that

$$0 = \Phi(T,0)\dot{\Phi}^{-1}(T,0)\bar{x} + \int_0^T \Phi(T,\tau)B(\tau)u(\tau)d\tau. \qquad (1.3.5)$$

Owing to (1.3.4), pre-multiplication of (1.3.5) by $\bar{x}'$ yields,

$$0 = \bar{x}'\bar{x} \qquad (1.3.6)$$

which is a contradiction.

We now particularize this result to the case in which A and B are constant, time-invariant matrices.

THEOREM 1.3.3 *If* A *and* B *are constant, then* (1.2.1) *is completely controllable if and only if*

$$\text{rank }[B,AB,\ldots,A^{n-1}B] = n. \qquad (1.3.7)$$

PROOF If the rank is less than n, there exists $\bar{x} \neq 0$ such that

$$B'(A')^i\bar{x} = 0, \quad i=0,\ldots,n-1. \qquad (1.3.8)$$

By the Cayley-Hamilton Theorem [2] the n-th and higher powers of A are linear combinations of the lower powers, so (1.3.8) is true for all i: consequently by Theorem 1.2.3(iv) we have

$$B'e^{A'(T-t)}\bar{x} = 0, \; t \in [0,T]. \qquad (1.3.9)$$

However the same theorem shows that in the case of constant matrices we have

$$\Phi(T,t) = e^{A(T-t)}, \qquad (1.3.10)$$

so that (1.3.9) implies that W(0,T) is singular, and hence that (1.2.1) is not completely controllable.

Suppose now that the rank is n, and assume that the system is not completely controllable. Then, from Theorem 1.3.2, W(0,T) is singular. Hence there is a non-zero vector $\bar{x}$ such that

$$B'(t)\Phi'(T,t)\bar{x} = 0, \quad t \in [0,T], \qquad (1.3.11)$$

which is

$$B' e^{A'(T-t)}\bar{x} = 0. \qquad (1.3.12)$$

Differentiating (n-1) times and setting t = T, we have

$$B'(A')^{i}\bar{x} = 0, \quad i=1,\ldots,n \qquad (1.3.13)$$

and this cannot hold unless the rank of $[B,AB,\ldots,A^{n-1}B]$ is less than n, a contradiction.

1.3.2 Observability

Here we consider the question of whether or not it is possible to de= termine the unknown initial state of the system

$$\dot{x}(t) = A(t)x(t) \qquad (1.3.14)$$

from a knowledge of the output y(t), $t \in [0,T]$, where

$$y(t) = C(t)x(t). \qquad (1.3.15)$$

DEFINITION 1.3.4 The pair (1.3.14-15) *is completely observable over the time interval* [0,T] *if the arbitrary initial state* x_0 *can be deter= mined from the function* y(t), $t \in [0,T]$, $T < \infty$.

Note that if u(t), $t \in [0,T]$ is known, observability of the pair (1.3.14-15) is the same as observability of the pair (1.2.1-2). Accordingly we consider only (1.3.14-15).

THEOREM 1.3.5 *The pair* (1.3.14-15) *is completely observable if and only if*

$$M(0,T) \triangleq \int_0^T \Phi'(\tau,0)C'(\tau)C(\tau)\Phi(\tau,0)d\tau \text{ is positive definite.} \qquad (1.3.16)$$

PROOF Since

$$y(t) = C(t)\Phi(t,0)x_0 \qquad (1.3.17)$$

we have

$$\int_0^T \Phi'(\tau,0)C'(\tau)y(\tau)d\tau = M(0,T)x_0. \qquad (1.3.18)$$

If M(0,T) is singular, there will, for a given function $y(\cdot)$ satisfying (1.3.15), be several solutions x_0 to (1.3.18); accordingly we should not have complete observability. On the other hand, if M(0,T) is non-singular and hence positive definite (it is always at least positive semi-definite), x_0 may be computed uniquely from (1.3.18).

As in the controllability case, the above theorem simplifies when A and C are constant matrices.

THEOREM 1.3.6 *If* A *and* C *are constant, then the pair* (1.3.14-15) *is completely observable if and only if*

$$\text{rank}\,[C',A'C',\ldots,(A')^{n-1}C'] = n. \qquad (1.3.19)$$

The similarity between Theorems 1.3.2 and 1.3.5 and Theorems 1.3.3 and 1.3.6 suggests the following duality between controllability and ob= servability.

THEOREM 1.3.7 *The pair*

$$\dot{x}(t) = A(t)x(t); \quad y(t) = C(t)x(t) \qquad (1.3.20)$$

is completely observable if and only if the system

$$\dot{z}(t) = -A'(t)z(t) + C'(t)v(t) \qquad (1.3.21)$$

is completely controllable.

1.3.3 Minimal realization

Consider an autonomous control system

$$\dot{x} = Ax + Bu, \quad y = Cx \qquad (1.3.22)$$

characterized by the matrix triple (A,B,C). By (1.2.4) and Theorem 1.2.3, the output resulting from input $u(\cdot)$ and zero initial state is

$$y(t) = \int_0^t Ce^{A(t-\tau)}Bu(\tau)d\tau$$
$$= \int_0^t T(t-\tau)u(\tau)d\tau.$$

DEFINITION 1.3.8 The matrix

$$T(t) \triangleq Ce^{At}B \qquad (1.3.23)$$

is called the weighting pattern of the system.

Clearly two systems which have the same weighting pattern have the same input-output behaviour, and this may even be true of systems with different state dimension. When (1.3.23) holds we say that the system (1.3.22) is a *realization* of the weighting pattern $T(\cdot)$, and the pro= blem arises of finding alternative realizations of $T(\cdot)$ having least possible state dimension. Such a realization is called a *minimal realization,* and the following theorem beautifully characterizes mini= mal realizations.

THEOREM 1.3.9 *The system (1.3.22) is a minimal realization of the weighting pattern (1.3.23) iff the system is both completely controllable and completely observable.*

A proof may be found in [3] , and related material is considered in Lecture 2.

1.3.4 Stability

We here particularize our system (1.2.1) to the case of no control and constant A matrix, viz.

$$\dot{x}(t) = Ax(t), \quad x(0) = x_0. \tag{1.3.24}$$

We define a stability matrix as follows.

DEFINITION 1.3.10 *A matrix A is a stability matrix if all of its eigenvalues have negative real parts.*

The following well-known theorem establishes the connection between A and the solution of (1.3.24).

THEOREM 1.3.11 [2] *A necessary and sufficient condition for the solution of (1.3.24) to approach zero as $t \to \infty$ (in this linear constant coefficient system this is asymptotic stability), regardless of the value of x_0, is that all the eigenvalues of A have negative real parts.*

A closely related theorem establishes the connection between A and the solution of a certain linear matrix equation.

THEOREM 1.3.12 [2] *A necessary and sufficient condition for A to be a stability matrix is that there exists a positive definite symmetric matrix S which satisfies the Liapunov equation*

$$SA + A^T S = -I. \tag{1.3.25}$$

The third lecture in this Series is devoted to the stability theory of ordinary differential equations, and Theorem 1.3.12 is a special case of Theorem 3.2.2.

1.3.5 Feedback

Many, though not all, of the approaches to the control of (1.2.1) yield linear expressions for the control $u(t)$ as a function of the system state $x(t)$. In other words, these approaches yield a matrix function of time $K(t)$ which synthesizes a control $u(t)$ according to

$$u(t) = -K(t)x(t). \tag{1.3.26}$$

For obvious reasons $K(t)$ is referred to as a 'feedback gain' matrix and (1.3.26) as a state 'feedback controller' for (1.2.1). In certain situations (which require A and B to be constant) K turns out to be constant, and then one is concerned with the question of whether or not the solutions of the 'closed loop' system

$$\dot{x}(t) = (A-BK)x(t)$$

go to zero as $t \to \infty$. According to Theorem 1.3.11, this occurs if and only if $A - BK$ is a stability matrix. The problem of whether or not there exists a matrix K such that $A-BK$ is a stability matrix, is called the stabilizability problem. Here we remark only that a sufficient condition for (1.2.1) to be stabilizable is that (1.2.1), with A and B constant, should be completely controllable. This will be proved in Lecture 2. Obviously this condition is not necessary, since if A is a stability matrix and $B \equiv 0$, we have that all solutions of (1.2.1) go to zero, but the system is clearly not completely controllable with this choice of A and B.

1.4 NEIGHBOURHOOD PROPERTIES OF CONTINUOUS NONLINEAR DYNAMIC SYSTEMS

In this section we state certain results on the controllability and stability of nonlinear systems in the neighbourhood of an operating point or an equilibrium point; the results depend heavily on those of Section 1.3 for linear systems, and accordingly illustrate the use=fulness of the theory of linear dynamic systems.

1.4.1 Controllability

We first introduce a definition of controllability suitable for (1.1.4).

DEFINITION 1.4.1 *The dynamic system* (1.1.4) *is completely null-control=lable if there exists an open set in* $\mathbb{R}^n$ *which contains the origin and for which any* x_0 *in this set can be steered to the origin at time* $T < \infty$ *by a continuous control function* $u(\cdot)$.

THEOREM 1.4.2 [4] *Consider the autonomous nonlinear system*

$$\dot{x}(t) = f(x(t),u(t)). \tag{1.4.1}$$

Assume

(i) $f(0,0) = 0$

(ii) $\text{rank}[B,AB,\ldots,A^{n-1}B] = n$

where

$$A \triangleq \frac{\partial f}{\partial x}(0,0), \quad B \triangleq \frac{\partial f}{\partial u}(0,0).$$

Then (1.4.1) *is completely null-controllable.*

Note that an analogous result holds in the case of observability [4]. Requirement (ii) of Theorem 1.4.2 is certainly not necessary, as the following example shows.

EXAMPLE 1.4.3

$$\dot{x} = u^3. \qquad (1.4.2)$$

This system is clearly completely null-controllable, but

$$A = \frac{\partial f}{\partial x}(0,0) = 0, \quad B = \frac{\partial f}{\partial u}(0,0) = 0$$

so that rank [B] = 0 < 1.

Despite the fact that Theorem 1.4.2 provides only a sufficient condi= tion for controllability in nonlinear dynamic systems, it is neverthe= less useful in many applications, and it emphasizes the role that linear system theory plays in deducing local properties of nonlinear systems.

1.4.2 Stability

As we shall deal with the autonomous nonlinear system

$$\dot{x}(t) = f(x(t)) \qquad (1.\ 4.3)$$

we first define precisely what we mean by stability.

DEFINITION 1.4.4 *The autonomous nonlinear dynamic system* (1.4.3) *is asymptotically stable about the origin if for each* $\varepsilon > 0$ *there is a* $\delta > 0$ *such that* $\|x_0\| < \delta$ *implies that* $x(t)$ *satisfies* $\|x(t)\| < \varepsilon$ *for all* $t \in [0,\infty)$, *and* $x(t) \to 0$ as $t \to \infty$.

Note that in the case of the linear dynamic system (1.3.24), the second condition implies the first.

We then have the following well-known theorem, which is proved in Lecture 3.

THEOREM 1.4.5 *Consider* (1.4.3) *and assume that*

(i) $f(0) = 0$;

(ii) $\frac{\partial f}{\partial x}(0)$ *is a stability matrix.*

Then (1.4.3) *is asymptotically stable about the origin.*

1.5. NOISE

In modelling real systems, uncertainty about the system or measurements is accounted for by the introduction of certain random variables. For example, the discrete-time linear dynamic system (1.2.11), (1.2.12) will become

$$x_{k+1} = A_k x_k + B_k u_k + \Gamma_k w_k \tag{1.5.1}$$

$$y_k = H_k x_k + v_k \tag{1.5.2}$$

where the sequences $\{w_k\}$, $\{v_k\}$ and the initial condition x_o are independent Gaussian random variables with the following statistics:

$$E[w_k] = 0 \quad , \quad E[w_j w_k'] = W_k \delta_{jk} \tag{1.5.3}$$

$$E[v_k] = 0 \quad , \quad E[v_j v_k'] = V_k \delta_{jk} \tag{1.5.4}$$

$$E[x_o] = \bar{x}_o \ , \quad E[(x_o - \bar{x}_o)(x_o - \bar{x}_o)'] = X_o . \tag{1.5.5}$$

Here E denotes expected value, and W_k, V_k, X_o are covariance matrices. The symbol δ_{jk} takes on the value unity if $j = k$, and is zero otherwise.

To describe completely a random sequence $\{x_o, x_1, \ldots, x_k\}$ it is in general necessary to specify the joint probability density

$$p[x_o, x_1, \ldots, x_k] , \tag{1.5.6}$$

but because the elements of the sequence $\{w_k\}$ are independent and independent of x_0 we see that (1.5.1) enjoys the Markov property, viz.

$$p[x_{k+1}|x_k,\ldots,x_0] = p[x_{k+1}|x_k]. \qquad (1.5.7)$$

Accordingly, the joint probability density function of the Markov pro= cess (1.5.1) is described completely by the initial density function $p[x_0]$ and the *transition density functions* $p[x_{k+1}|x_k]$.

The sequences $\{w_k\}$ and $\{v_k\}$ can be thought of as *purely random sequences* because, for example,

$$p[w_{k+1}|w_k,\ldots,w_0] = p[w_{k+1}] \text{ for all } k. \qquad (1.5.8)$$

In certain applications it is appropriate for the Gaussian sequence $\{w_k\}$ and/or $\{v_k\}$ and/or x_0 to be correlated, but such correlated noises can themselves be modelled adequately by feeding purely random Gaussian sequences into a suitable discrete-time dynamic system. Accordingly, there is in this respect no loss of generality in restricting one's attention to models of the form (1.5.1)-(1.5.5).

In continuous time, owing to fundamental mathematical difficulties, the situation is more awkward, and we shall consequently not consider this case here. Rather, we refer the reader to [5] for an introductory treatment. When in later lectures we extend certain results from dis= crete time to continuous time, we shall treat the continuous-time noisy case formally and not rigorously.

As indicated above, it is common to assume Gaussian density functions for the noises because these densities reproduce (i.e. remain Gaussian) under linear transformations. As we shall see later, under some cir= cumstances this assumption is inessential and can be dispensed with.

1.6 REFERENCES AND NOTES

[1] E.A. Coddington and N. Levinson, *Theory of ordinary differential equations*. McGraw-Hill, New York, 1955.

[2] R.E. Bellman, *Introduction to matrix analysis*. McGraw-Hill, New York, 1970.

[3] R.W. Brockett, *Finite dimensional linear systems*. Wiley, New York, 1970.

[4] E.B. Lee and L. Markus, *Foundations of optimal control theory*. Wiley, New York, 1967.

[5] K.J. Astrom, *Introduction to stochastic control theory*. Academic Press, New York, 1970.

The foundations of the modern state space theory of linear control systems were laid by R.E. Kalman in a series of famous papers commen= cing with: Contributions to the theory of optimal control, *Bol. Soc. Mat. Mexicana* 5 pp 102-119, 1960; and On the general theory of control systems, *Proc. 1st IFAC Congress*, Moscow, Butterworths, London. A very readable introduction to the theory, written by Kalman, forms Chapter 2 of the book *Topics in Mathematical System Theory* by R.E. Kalman, P.L. Falb and M.A. Arbib, published by McGraw-Hill Book Company, New York, 1969. Useful books include the following texts.

C.T. Chen, *Introduction to linear system theory*. Holt, Rinehart and Winston, New York, 1970.

J.P. LaSalle and S. Lefschetz, *Stability by Liapunov's Direct Method*. Academic Press, New York, 1961.

J.S. Meditch, *Stochastic optimal linear estimation and control*. McGraw-Hill, New York, 1969.

J.E. Rubio, *The theory of linear systems*. Academic Press, New York, 1971.

2. CANONICAL FORMS, POLE ASSIGNMENT AND STATE OBSERVERS

2.1 CANONICAL FORMS

2.1.1 Equivalent systems

This lecture briefly surveys some standard material on the structure and regulation of autonomous linear control systems

$$\dot{x} = Ax + Bu \,, \qquad y = Cx \tag{2.1.1}$$

maintaining for the present a deterministic viewpoint which does not take into account random perturbations to u, x or y.

If $x(t) = 0$, for $t \leqslant 0$, then from the variation of constants formula (1.2.4) we have the *input-output* relation

$$y(t) = \int_0^t Ce^{A(t-\tau)}Bu(\tau)d\tau$$

involving the matrix function

$$Ce^{At}B,$$

which is known as the *weighting pattern* of the system. Alternatively, denoting the Laplace transform of $u(t)$ by $\bar{u}(s)$ etc., we have

$$s\bar{x}(s) = A\bar{x}(s) + B\bar{u}(s),$$

whence

$$\bar{y}(s) = C(sI-A)^{-1}B\bar{u}(s).$$

The matrix function

$$C(sI-A)^{-1}B$$

is the Laplace transform form of the weighting pattern, and is called the *transfer matrix* of the system. Both of these matrix functions describe the input-output behaviour of the system directly, and it turns out that they are invariant under arbitrary invertible transfor= mations of the state variables. Thus, if S is any invertible n×n matrix, and we introduce new state variables,

$$\tilde{x} \overset{\Delta}{\equiv} Sx \text{ so that } x = S^{-1}\tilde{x}, \qquad (2.1.2)$$

then (2.1.1) is equivalently given by

$$\dot{\tilde{x}} = SAS^{-1}\tilde{x} + SBu$$
$$y = CS^{-1}\tilde{x}.$$

Thus our matrix triple (A,B,C) is transformed into $(\tilde{A},\tilde{B},\tilde{C}) = (SAS^{-1}, SB, CS^{-1})$,

and it is easily checked that, as remarked above, we have

$$\tilde{C}e^{\tilde{A}t}\tilde{B} \equiv Ce^{At}B \text{ and } \tilde{C}(sI-\tilde{A})^{-1}\tilde{B} \equiv C(sI-A)^{-1}B.$$

2.1.2 The general structure theorem

This invariance suggests that we might well make use of suitable state transformation matrices S to bring about both simplification and clari= fication of the structure of the system.

The first result of this type essentially decomposes a general system into four distinct subsystems (some of which may be absent)

THEOREM 2.1.1 *Under a suitable transformation* (2.1.2) *every system* (2.1.1) *is equivalent to one of the form*

$$\begin{bmatrix} \dot{\tilde{x}}_1 \\ \dot{\tilde{x}}_2 \\ \dot{\tilde{x}}_3 \\ \dot{\tilde{x}}_4 \end{bmatrix} = \begin{bmatrix} \tilde{A}_{11} & \tilde{A}_{12} & \tilde{A}_{13} & \tilde{A}_{14} \\ 0 & \tilde{A}_{22} & 0 & \tilde{A}_{24} \\ 0 & 0 & \tilde{A}_{33} & \tilde{A}_{34} \\ 0 & 0 & 0 & \tilde{A}_{44} \end{bmatrix} \begin{bmatrix} \tilde{x}_1 \\ \tilde{x}_2 \\ \tilde{x}_3 \\ \tilde{x}_4 \end{bmatrix} + \begin{bmatrix} \tilde{B}_1 \\ \tilde{B}_2 \\ 0 \\ 0 \end{bmatrix} u \qquad (2.1.3)$$

$$y = [0 \mid \tilde{C}_2 \mid 0 \mid \tilde{C}_4] \quad [\tilde{x}_1 \mid \tilde{x}_2 \mid \tilde{x}_3 \mid \tilde{x}_4]'$$

in which the (1,2) *subsystem is completely controllable and the* (2,4) *subsystem is completely observable.*

PROOF Referring to the system (2.1.1), let U and V be the controlla= bility and observability matrices

$$U = [B \mid AB \mid A^2B \mid \ldots \mid A^{n-1}B] \;, \quad V = [C' \mid A'C' \mid \ldots \mid A'^{n-1}C']$$

respectively, and let

$$X = \text{range } U \quad , \quad Y = \ker V'.$$

These subspaces of $\mathbb{R}^n$ are invariant under the system matrix A. For example, if $v \in Y$, then

$$Cv = CAv = CA^2v = \ldots = CA^{n-1}v = 0. \qquad (2.1.4)$$

The last n-1 of these equations may be written as

$$C(Av) = CA(Av) = \ldots = CA^{n-2}(Av) = 0.$$

To show that also $CA^{n-1}(Av) = CA^nv = 0$, we use the famous Cayley-Hamilton theorem, according to which if

$$|\lambda I-A| \equiv \lambda^n + a_1\lambda^{n-1} + a_2\lambda^{n-2} + \ldots + a_{n-1}\lambda + a_n$$

is the characteristic polynomial of A, then

$$A^n + a_1A^{n-1} + a_2A^{n-2} + \dots + a_{n-1}A + a_nI = 0.$$

It follows that CA^n is a linear combination of $C, CA, \dots, CA^{n-1}$, and hence that (2.1.4) implies also that $CA^n v = 0$. Thus if $v \in Y$, then $Av \in Y$. We can show similarly that $AX \subseteq X$.

We now select a basis

$$a_1,\dots,a_p,b_1,\dots,b_q,c_1,\dots,c_r,d_1,\dots,d_s$$

(with $p+q+r+s = n$) for $\mathbb{R}^n$ as follows: let $a_1,\dots,a_p$ be a basis for $X \cap Y$. This may be augmented to give a basis $a_1,\dots,a_p,b_1,\dots,b_q$ for X, and likewise augmented to give a basis $a_1,\dots,a_p,c_1,\dots,c_r$ for Y; the vectors a_i,b_j,c_k together being then a basis for the direct sum $X + Y$. Finally augment this latter basis to produce the desired basis for $\mathbb{R}^n$.

Let S be that invertible matrix such that

$$S^{-1} = [a_1 \vdots \dots \vdots a_p \vdots b_1 \vdots \dots \vdots b_q \vdots c_1 \vdots \dots \vdots c_r \vdots d_1 \vdots \dots \vdots d_s],$$

and denote the *rows* of S, in order by

$$\alpha_1',\dots,\alpha_p',\beta_1',\dots,\beta_q',\gamma_1',\dots,\gamma_r',\delta_1',\dots,\delta_s'$$

Then from the relation $SS^{-1} = I$ we have, inter alia, for any indices i,j in the appropriate ranges,

$$\gamma_i'a_j = 0, \quad \gamma_i'b_j = 0, \quad \delta_i'a_j = 0, \quad \delta_i'b_j = 0 \qquad (2.1.5)$$

Since $AX \subseteq X$, each of the vectors Aa_j and Ab_j is a linear combination of the basis vectors $a_1,\dots,a_p,b_1,\dots,b_q$, and hence it follows from the above relations that

$$\gamma_i' Aa_j = 0, \quad \gamma_i'Ab_j = 0, \quad \delta_i'Aa_j = 0, \quad \delta_i'Ab_j = 0.$$

Since, for example, $\gamma_i' A a_j$ is the i/j entry of the 3/1 block of the matrix $\tilde{A} = SAS^{-1}$, these relations mean precisely that the 3/1, 3/2, 4/1 and 4/2 blocks of $\tilde{A}$ are zero. It follows similarly from the in= variance of Y under A that the 2/1, 2/3, 4/1 and 4/3 blocks of $\tilde{A}$ are zero. Turning to $\tilde{B} = SB$, since range $B \subseteq$ range $X = \text{span}\{a_1, \ldots, b_q\}$, it follows from (2.1.5) that the two lower blocks of $\tilde{B}$ vanish. The two zero blocks of $\tilde{C} = CS^{-1}$ arise similarly.

This theorem is interpreted as follows. The state-space is decomposed into four direct summands, in which $\tilde{x}_2$ is both c.c. and c.o., $\tilde{x}_1$ is c.c. but unobservable, $\tilde{x}_3$ is uncontrollable and unobservable, while $\tilde{x}_4$ is uncontrollable but is c.o. Using the series expansion for e^{At} it is easily verified that the weighting pattern for (2.1.3) reduces to $C_2 e^{A_{22}t} B_2$. Owing to the invariance of the weighting pattern it follows that the original system (2.1.1) has the same weighting pattern as the completely controllable and completely observable sub-system

$$\dot{\tilde{x}}_2 = \tilde{A}_{22}\tilde{x}_2 + \tilde{B}_2 u \ , \quad y = \tilde{C}_2 x_2 .$$

In fact, by Theorem 1.3.9, this subsystem is a *minimal* realization of the weighting pattern of the original system.

2.1.3 A canonical form for completely controllable systems

Consider the n-th order differential equation (with $D = \frac{d}{dt}$)

$$(D^n + \alpha_n D^{n-1} + \ldots + \alpha_2 D + \alpha_1)y = u(t) \qquad (2.1.6)$$

This may be regarded as defining an input-output system with input u(·) and output y(·) depending upon u(·) and upon initial conditions for y and its first n-1 derivatives. This equation is most simply reduced to the standard form (2.1.1) by introducing state variables

$$x_1 = y, \quad x_2 = D_y = \dot{x}_1 , \quad x_3 = D^2{}_y = \dot{x}_2, \ldots, x_n = D^{n-1}{}_y = \dot{x}_{n-1},$$

so that (2.1.6) becomes

$$\dot{x}_n + \alpha_n x_n + \alpha_{n-1} x_{n-1} + \ldots + \alpha_1 x_1 = u.$$

In matrix form these equations are

$$\begin{pmatrix} \dot{x}_1 \\ \dot{x}_2 \\ \vdots \\ \dot{x}_n \end{pmatrix} = \begin{pmatrix} 0 & 1 & 0 & \ldots & 0 & 0 \\ 0 & 0 & 1 & & 0 & 0 \\ & & & \ddots & & \\ 0 & 0 & 0 & \ldots & 0 & 1 \\ -\alpha_1 & -\alpha_2 & -\alpha_3 & \ldots & -\alpha_{n-1} & -\alpha_n \end{pmatrix} \begin{pmatrix} x_1 \\ x_2 \\ \vdots \\ x_n \end{pmatrix} + \begin{pmatrix} 0 \\ 0 \\ \vdots \\ 0 \\ 1 \end{pmatrix} u \triangleq \bar{A}x + \bar{b}u. \quad (2.1.7)$$

$$y = [1,0,\ldots,0]x. \quad (2.1.8)$$

Here the coefficient matrix $\bar{A}$ is in so-called *companion form,* and it is easily checked that the controllability matrix

$$\bar{U} = [\bar{b} \,\vdots\, \bar{A}\bar{b} \,\vdots\, \bar{A}^2\bar{b} \,\vdots\, \ldots \,\vdots\, \bar{A}^{n-1}\bar{b}]$$

is lower-cross-triangular with 1's on the cross diagonal, and is there= fore invertible. Thus (2.1.7) is completely controllable. For (2.1.6) this means that given any two vectors $x^{(1)}$, $x^{(2)}$ of prescribed values for y and its first n-1 derivatives, and any time interval [0,T], there is a control $u(\cdot)$ on [0,T] which will produce the values $x^{(2)}$ at time t=T from the values $x^{(1)}$ at time t = 0. Note also that the system poles (i.e. the eigenvalues of $\bar{A}$) are precisely the zeros of the charac= teristic polynomial

$$\lambda^n + \alpha_n \lambda^{n-1} + \ldots + \alpha_2 \lambda + \alpha_1 = 0$$

of (2.1.6) - thus the characteristic polynomial of a companion matrix can be read off from its bottom row.

The special form (2.1.7) is of interest because it is in fact a canoni= cal form for completely controllable single-input systems - i.e. any such system (2.1.1) is equivalent under a suitable transformation (2.1.2) to one of the form (2.1.7): the required matrix S^{-1} has columns v_i generated recursively according to the formulas

$$v_n = B, \quad v_{k-1} = Av_k + \alpha_k B \quad k=n,n-1,\ldots,2.$$

We leave details of verification to the reader.

For multiple-input systems this situation generalizes as follows. Imagine given r ordinary differential equations of the form

$$\sum_{j=1}^{r} P_{ij}(D)\, y_j = u_i(t) \quad i=1,2,\ldots,r \qquad (2.1.9)$$

having r input functions $u_i(t)$ and r outputs $y_i(t)$. Here the P_{ij} are monic polynomials in $D = d/dt$, such that in the i-th equation there are higher derivatives in y_i than in other outputs:

$$n_i \triangleq \deg P_{ii} > \deg P_{ij} \text{ for } j \neq i.$$

Then defining

$$x_j^{(i)} = D^{j-1} y_i \quad \text{for } j=1,2,\ldots,n_i; \quad i=1,\ldots,r$$

and arranging these into the state vector

$$x' = [x_1^{(1)},\ldots,x_{n_1}^{(1)},x_1^{(2)},\ldots,x_{n_2}^{(2)},\ldots,x_1^{(r)},\ldots,x_{n_r}^{(r)}],$$

we find that (2.1.9) takes the form (2.1.1) with

$$\dot{x} = \left(\begin{array}{c|c|c|c|c} \Gamma_{n_1} & 0 & 0 & \cdots\cdots & 0 \\ \hline \multicolumn{5}{c}{n_1\text{-th row full}} \\ \hline 0 & \Gamma_{n_2} & 0 & \cdots\cdots & 0 \\ \hline \multicolumn{5}{c}{(n_1+n_2)\text{-th row full}} \\ \hline 0 & 0 & \Gamma_{n_3} & \cdots\cdots & 0 \\ \hline \vdots & \vdots & \vdots & \ddots & \vdots \\ \hline 0 & 0 & 0 & \cdots\cdots & \Gamma_{n_r} \\ \hline \multicolumn{5}{c}{\text{last row full}} \\ \hline \end{array}\right) x + \left(\begin{array}{c} 0 \\ \hline 10.0 \\ \hline 0 \\ \hline 010.0 \\ \hline 0 \\ \vdots \\ 0 \\ \hline 0..01 \end{array}\right) u ,$$

where for any integer k > 1, Γ_k is the (k-1) x k matrix

$$\Gamma_k = \begin{pmatrix} 0 & 1 & 0 & \dots & 0 \\ 0 & 0 & 1 & \dots & 0 \\ \\ 0 & 0 & 0 & .. 0 & 1 \end{pmatrix} .$$

Thus the A-matrix is almost block-companion - this is spoiled only by the non-zero entries in the n_1-th, (n_1+n_2)-th,... rows. Note that for r = 1 this reduces to (2.1.7).

The following theorem is due to Luenberger, and its proof is a clever extension of that sketched above for single-input systems.

THEOREM 2.1.2 *Given any completely controllable system* (2.1.1) *with state dimension* n *and input dimension* $r \leq n$ *with rank* $B = r$, *there exists a state transformation* (2.1.2) *and a non-singular transformation* T *in control space* $\mathbb{R}^r$, *such that with new state variables and control variables*

$$\tilde{x} = Sx, \qquad \tilde{u} = Tu$$

the matrices

$$\tilde{A} = SAS^{-1} \quad , \qquad \tilde{B} = SBT^{-1}$$

take the form exhibited in (2.1.10). *The integers* $n_1, n_2 \ldots, n_r$ *are uniquely fixed by the pair* (A,B).

We shall apply these theorems in the next section to establish an im= portant result of practical value. Note also that, using duality, counterparts for completely observable systems follow.

2.2 STABILIZABILITY AND POLE PLACEMENT

One of the most important questions in the design of regulators is the stabilizability problem : for a given system (2.1.1), can we find a linear feedback law

$$u = -Kx$$

which stabilizes the system, i.e. such that the 'closed-loop' or 'self-controlling' system

$$\dot{x} = (A-BK)x \qquad (2.2.1)$$

is asymptotically stable? Mathematically this is the question: given n x n and n x r real matrices A,B, can one find an r x n matrix K such that the eigenvalues of A-BK all have negative real parts? If so, we say that the pair (A,B) is *stabilizable*.

A considerably stronger question is the following : given any real monic polynomial

$$P(\lambda) = \lambda^n + a_n\lambda^{n-1} + a_{n-1}\lambda^{n-2} + \ldots + a_2\lambda + a_1,$$

can we find an r x n matrix K such that $P(\lambda)$ is the characteristic polynomial of A-BK? This is the *pole placement problem*, for it is equivalent to asking whether we can place the poles of the closed loop system (2.2.1) wherever we like (subject of course to complex poles appearing in conjugate pairs) by suitable choice of the gain matrix K.

THEOREM 2.2.1 *For a given system (2.1.2) the closed-loop poles may be arbitrarily placed iff the system is completely controllable. If*

$$\text{rank } [B, AB, \ldots, A^{n-1}B] = n_c < n,$$

then a certain group of $n-n_c$ *of the eigenvalues of* A *are eigenvalues of* A-BK, *regardless of the choice of* K, *while the remaining* n_c *eigenvalues of* A-BK *may be arbitrarily placed (subject to conjugacy requirements).*

PROOF If $n_c < n$, we first apply the controllability part of Theorem 2.1.1 to obtain an equivalent system with A and B in the form

$$\bar{A} = \begin{pmatrix} \bar{A}_1 & \bar{A}_2 \\ 0 & \bar{A}_3 \end{pmatrix} \qquad \bar{B} = \begin{pmatrix} \bar{B}_1 \\ 0 \end{pmatrix},$$

where $\bar{A}_1$ is $n_c \times n_c$ and the pair $(\bar{A}_1, \bar{B}_1)$ is completely controllable. If $n_c = n$, the original system is completely controllable, and the matrices $\bar{A}_2$, $\bar{A}_3$ are vacuous. Theorem 2.1.2 then ensures that we may carry out a further transformation, affecting only the first n_c compo= nents of the state vector, and also the control vector, to produce an equivalent system

$$\tilde{A} = \begin{pmatrix} \tilde{A}_1 & \tilde{A}_2 \\ 0 & \tilde{A}_3 \end{pmatrix} \qquad \tilde{B} = \begin{pmatrix} \tilde{B}_1 \\ 0 \end{pmatrix}$$

with $\tilde{A}_1$ and $\tilde{B}_1$ as in Luenberger's canonical form (2.1.10).

For any gain matrix $\tilde{K} = [\tilde{K}_1, \tilde{K}_2]$ we have

$$\tilde{A}-\tilde{B}\tilde{K} = \left(\begin{array}{c|c} \tilde{A}_1-\tilde{B}_1\tilde{K}_1 & \tilde{A}_2-\tilde{B}_1\tilde{K}_2 \\ \hline 0 & \tilde{A}_3 \end{array}\right),$$

so that the characteristic polynomial of $\tilde{A}-\tilde{B}\tilde{K}$ is the product of the factor

$$\det\,[\lambda I - \tilde{A}_3]$$

of degree $n-n_c$ independent of the choice of $\tilde{K}$, and the factor

$$\det\,[\lambda I - \tilde{A}_1+\tilde{B}_1\tilde{K}_1]$$

of degree n_c. Since

$$\tilde{B}_1\tilde{K}_1 = \begin{bmatrix} 0 \\ \hline 10..0 \\ \hline 0 \\ \hline 010.0 \\ \hline \vdots \\ 0 \\ \hline 0...01 \end{bmatrix} \begin{bmatrix} \tilde{k}_{11} & \tilde{k}_{12} \cdots & \tilde{k}_{1n} \\ \tilde{k}_{r1} & \tilde{k}_{r2} \cdots & \tilde{k}_{rn} \end{bmatrix} = \begin{bmatrix} 0 \\ \hline \tilde{k}_{11} \;\cdots\; \tilde{k}_{1n} \\ \hline 0 \\ \hline \tilde{k}_{21} \;\cdots\; \tilde{k}_{2n} \\ \hline 0 \\ 0 \\ \hline \tilde{k}_{r1} \;\cdots\; \tilde{k}_{rn} \end{bmatrix},$$

it is clear from (2.1.10) that the matrix $\tilde{A}_1-\tilde{B}_1\tilde{K}_1$ has the same form as A_1, with the elements $a_{n_i j}$ of the full rows being modified by the amount $-\tilde{k}_{ij}$. Thus by adjusting the choice of $\tilde{K}_1$ these full rows of $\tilde{A}_1-\tilde{B}_1\tilde{K}_1$ can be arbitrarily set. In particular we may choose $\tilde{K}_1$ so that $\tilde{A}_1-\tilde{B}_1\tilde{K}_1$ becomes the $n_c \times n_c$ companion matrix corresponding to any real monic polynomial of degree n_c. The proof is completed by returning to the original state variables, and noting that the corresponding matrix A-BK is similar to $\tilde{A}-\tilde{B}\tilde{K}$, and hence has the same characteristic polyno= mial.

2.3 STATE OBSERVERS AND THE AUGMENTED SYSTEM

2.3.1 State observers

In order to implement a *state feedback law*

$$u = -Kx \tag{2.3.1}$$

it is essential to be able to continually measure each state variable, and often this is impracticable or even impossible. Let us suppose that we can measure only $m < n$ of the state variables, or more gene= rally, that m independent linear combinations of the state variables can be measured. Then there is an $m \times n$ matrix C such that each com= ponent of the output

$$y = Cx$$

can be measured. Then we might hope that an *output feedback* law

$$u = -Ky$$

might result in a stable closed-loop system

$$\dot{x} = (A-BKC)x \tag{2.3.2}$$

However if $r + m$ is small compared with n, there are too few entries in the $r \times m$ matrix K to achieve much by way of pole placement.

THEOREM 2.3.1 Let $\nu = \min(n,r+m-1)$. *For almost all*[1] *completely controllable and observable systems* (A,B,C), ν *poles of the closed loop system* (2.3.2) *can be placed arbitrarily close to the zeros of any* ν*-th degree real polynomial by suitable choice of* K.

If $\nu < n$, this is not much use, and it becomes necessary to resort to a *dynamic* feedback law involving what is known as a *state observer* . This is itself a linear system which must be designed and built, and which operates alongside and interconnected with the given system. Its inputs are the output of the given system together with a copy of the

---

1) A certain matrix must have maximal rank - hence this phrase 'almost all'.

input to the original system, and it should output an estimate z(t) of the full state vector x(t) of the original system.

DEFINITION 2.3.2 Let

$$\dot{x} = Ax + Bu, \quad y = Cx \tag{2.3.3}$$

be a given system. A state observer for (2.3.3) *is a linear system*

$$\dot{z} = Fz + Gu + Hy, \quad z \in \mathbb{R}^n \tag{2.3.4}$$

with the property that, for any initial state x_0 *and initial observer state* z_0*, and any controller* $u(\cdot)$*, the solution* $(x(t), z(t))$ *of* (2.3.3) ∪ (2.3.4) *satisfies*

$$z(t) - x(t) \to 0 \text{ as } t \to \infty$$

Mathematically, observer design is the problem of the choice of the matrices F,G,H. The observer error is

$$e(t) = z(t) - x(t),$$

and we have

$$\dot{e} = \dot{z} - \dot{x} = F(e+x) + Gu + HCx - Ax - Bu$$

or

$$\dot{e} = Fe + (F-A+HC)x + (G-B)u.$$

Thus the error dynamics will be independent of $u(\cdot)$ (and its accumulated effect on x) if we choose F,G,H to satisfy

$$G = B, \quad F = A - HC;$$

for then

$$\dot{e} = Fe. \tag{2.3.5}$$

Then to ensure that $e(t) \to 0$ as $t \to \infty$, we must at the same time ensure that F is a stability matrix. Thus we are faced with the problem of so choosing H that

$$F = A - HC$$

is a stability matrix.

Since F' has the same eigenvalues as F, this is equivalent to the problem of choosing H' such that F' = A' - C'H' is a stability matrix, or better still, such that F' has prescribed stable eigenvalues. This problem is mathematically identical with the pole assignment problem treated in Section 2.3.2 - only A is replaced by A' and B by C'. In particular, by Theorem 2.2.1, *if*

$$\text{rank}\ [C' \mid A'C' \mid \dots \mid A'^{n-1}C'] = n,$$

i.e. if the given system (2.3.3) *is completely observable, then given any n-th degree polynomial* $P(\lambda)$, *there exists a state observer* (2.3.4) *for which the poles of the observation error system* (2.3.5) *(i.e. the eigenvalues of* F*) are precisely the zeros of* $P(\lambda)$.

2.3.2 The augmented closed loop system

Suppose that we have selected a feedback law

$$u = -Kx \qquad (2.3.6)$$

for the system (2.1.1) and that, being able to measure only the components of the output vector y, we have built a state observer (2.3.4) which outputs an estimate z of the state x. The idea is then of course to replace the unusable feedback law (2.3.6) by

$$u = -Kz. \qquad (2.3.7)$$

resulting in the 2n-dimensional *augmented system*

$$\begin{pmatrix} \dot{x} \\ \dot{z} \end{pmatrix} = \begin{pmatrix} A & -BK \\ HC & F-GK \end{pmatrix} \begin{pmatrix} x \\ z \end{pmatrix} .$$

If we carry out the augmented state transformation

$$\begin{pmatrix} x \\ z \end{pmatrix} \rightarrow \begin{pmatrix} x \\ e \end{pmatrix} = \begin{pmatrix} x \\ z-x \end{pmatrix} = \begin{pmatrix} I & 0 \\ -I & I \end{pmatrix} \begin{pmatrix} x \\ z \end{pmatrix},$$

this system transforms into the equivalent one:

$$\begin{pmatrix} \dot{x} \\ \dot{e} \end{pmatrix} = \begin{pmatrix} A - BK & - BK \\ 0 & A-HC \end{pmatrix} \begin{pmatrix} x \\ e \end{pmatrix} ,$$

from which it is evident that the poles of the augmented system are the eigenvalues of the desired closed loop system matrix A-BK together with the eigenvalues of the observation error system matrix F = A - HC. By way of summary we state the following theorem.

Theorem 2.3.3 Suppose the system

$$\dot{x} = Ax + Bu, \quad y = Cx \qquad (2.3.8)$$

is both completely controllable and completely observable, and let $P(\lambda)$ *be any stable 2n-th degree real polynomial. There exist matrices* K,H *such that the system*

$$\dot{z} = (A - HC)z + Bu + Hy$$

is a state observer for (2.3.8) *and such that the feedback law*

$$u = -Kz$$

results in an asymptotically stable augmented system the poles of which are precisely the zeros of $P(\lambda)$.

2.3.3 Concluding remarks

This design procedure for a dynamic feedback controller is robust in the sense that if one has slightly incorrect values for the system matrices A,B,C, the augmented system should still be asymptotically stable. This is because eigenvalues depend continuously upon matrix elements.

It can be shown that when there are m independent observable outputs, the observer need only have dimension n-m. However, the mathematical treatment of such reduced-order observers is more complicated, and in any case, when one admits system and measurement noise into the model, a full n-th order observer (then called a state estimator, of which the Kalman filter is the commonest design) is required (see Lecture 5 below).

2.4 HISTORICAL NOTES AND REFERENCES

2.1 The general structure theorem 2.1.1 is due to R.E. Kalman, Canonical structure of linear dynamical systems. *Proc. Nat. Acad. Sci. (U.S.A.)*,48,pp 596-600, 1962,and L. Weiss and R.E. Kalman, Contributions to linear system theory, *Int. J. Engr. Sci.*,3, pp 141-171. 1965. The canonical form given in Theorem 2.1.2 is one of several for completely controllable systems given by D.G. Luenberger, Canonical Forms for linear multi-variable systems. *IEEE Trans. Aut. Control* AC-12,pp 290-293 (1967).See also the recent survey articles:

Maroulas, J. and S. Barnett, Canonical forms for time-invariant linear control systems: a survey with extensions, Part I. Single-input case. *Int. J. Systems Sci.*, 9, pp. 497-514, 1978; Part II. Multivariable Case, *Int. J. Systems Sci.*, 10, pp 33-50, 1979.

2.2 For single-input systems J.E. Betram in 1959 first proved that complete controllability implies that closed loop-poles can be freely positioned. For $r > 1$ the result is due to W.M. Wonham. On pole assignment in multi-input controllable linear systems. *IEEE Trans. Aut. Control*. AC-12, pp 660-665, (1967). The proof given here, based upon Luenberger's canonical form, is due to M. Heymann, Comments on 'On pole assignment in multi-input con= trollable linear systems.' *IEEE Trans Aut.Control*, AC-13, p 748, (1968).

2.3 Observers having state dimension less than n were first introduced by D.G. Luenberger in two papers : Observing the state of a linear system. *IEEE Trans. Military Electronics*, MIL-8, pp 74-80, 1964, and Observers for multivariable systems. *IEEE Trans. Aut. Control*, AC-11, pp 190-197, 1966. See also his paper: An introduc= tion to observers, *IEEE Trans. Aut. Control*, AC-16, pp 596-602, 1971.

Theorem 2.3.3 and related matters are treated in the following papers: H. Kimura, Pole assignment by gain output feedback. *IEEE Trans. Aut. Control*. AC-20, pp 509-516, 1975.

E.J. Davison, S.H. Wang, On pole assignment in linear multivariable systems using output feedback. *IEEE Trans. Aut. Control*, AC-20, pp 516-518, 1975.

The topics dealt with in this lecture are treated in many texts, including the following.

S. Barnett, *Introduction to Mathematical Control Theory*, Clarendon Press, Oxford, 1975.

R.W. Brockett, *Finite Dimensional Linear Systems*. John Wiley and Sons, Inc., New York, 1970.

R.E. Kalman, P.L. Falb and M.A. Arbib, *Topics in Mathematical System Theory*. McGraw-Hill Book Company, New York, 1969.

W.M. Wonham, *Linear Multivariable Control*. Springer-Verlag, New York, 1974.

3. LYAPUNOV STABILITY THEORY

3.1 GENERAL THEORY

3.1.1 Definitions and concepts

Let

$$\dot{x} = \frac{dx}{dt} = f(x), \quad t \geq 0 \qquad (3.1.1)$$

be an autonomous system of differential equations, with $x \in \mathbb{R}^n$, and $f: \mathbb{R}^n \to \mathbb{R}^n$ being continuously differentiable. More generally, x may represent an arbitrary point on a differentiable manifold M (being, for example, the phase space of a dynamical system with equations of motion (3.1.1)), with f being a smooth vector field on M. A point $\hat{x}$ at which

$$f(\hat{x}) = 0$$

is called a *rest point* (or equilibrium point) of the system (3.1.1).

Elementary stability theory concerns the behaviour of solutions of (3.1.1) which start at t = 0 near to a rest point $\hat{x}$: do all the solutions remain near $\hat{x}$, perhaps even tending to $\hat{x}$ as $t \to \infty$, or are there solutions starting arbitrarily near $\hat{x}$ which leave some neighbourhood of $\hat{x}$? In the former case we should say that the rest point is stable, or even asymptotically stable, while in the latter case it is unstable. The following definitions, first explicitly given by the Russian mathematician A.M. Lyapunov in 1892, are rigorous statements of the above ideas.

DEFINITION 3.1.1 A rest point $\hat{x}$ of the system (3.1.1) *is stable (in the sense of Lyapunov) if given any bounded neighbourhood*[1] *U of $\hat{x}$, there*

---

1) The term *neighbourhood* is used to mean 'open neighbourhood'. A neighbourhood is *bounded* iff its closure is compact.

exists a second neighbourhood $V \subseteq U$ of $\hat{x}$ *such that any solution* $x(\cdot)$ *of* (3.1.1) *which starts in* V at $t = 0$ *never leaves* U *(and hence is defined for all* $t \geqslant 0$*)*: i.e.

$$x(0) \in V \text{ implies } x(t) \in U \text{ for all } t \geqslant 0.$$

If, in addition, $x(0) \in V$ *implies that*

$$x(t) \to \hat{x} \text{ as } t \to \infty,$$

the rest point $\hat{x}$ *is said to be asymptotically stable.*

If a rest point is stable, but not asymptotically so, we say that it is *merely stable*. The term *unstable* means simply 'not stable'.

3.1.2 Examples and counter-examples

(a) In $\mathbb{R}^1$, for the linear equation

$$\dot{x} = ax, \ t \geqslant 0$$

with solution

$$x(t) = x(0)e^{at},$$

the origin is a rest point which is asymptotically stable if $a < 0$, merely stable for $a = 0$, and unstable for $a > 0$.

(b) The damped harmonic oscillator equation

$$\ddot{x} + 2\alpha\dot{x} + x = 0,$$

with damping coefficient $\alpha \geqslant 0$, can be put in the form

$$\dot{x}_1 = x_2$$

$$\dot{x}_2 = -x_1 - 2\alpha x_2,$$

in which the system matrix has eigenvalues

$$-\alpha \pm \sqrt{\alpha^2-1}.$$

These are both negative or have negative real parts, except when $\alpha = 0$, from which it is clear that the origin $(x_1,x_2) = (0,0)$ is an asymptotically stable rest point if $\alpha > 0$, but is merely stable in the absence of damping.

(c) Let $q = \hat{q}$ denote an equilibrium configuration of a conservative autonomous mechanical system - i.e. if $V(q)$ is the potential energy function, then $(\nabla V)(\hat{q}) = 0$. If $\hat{q}$ is actually a strict local minimizer for V, then, in phase space, the state $(q,\dot{q}) = (\hat{q},0)$ is a merely stable rest point of the equations of motion. Historically, this is the original example! The proof is given later.

(d) One might be tempted into thinking that a rest point $\hat{x}$ of (3.1.1) must be asymptotically stable if all solutions starting in some neighbourhood of $\hat{x}$ are defined for all $t \geqslant 0$ and tend to $\hat{x}$. A simple counter-example, involving a differential equation on a circle, is given by

$$\dot{\theta} = 1 - \cos\theta$$

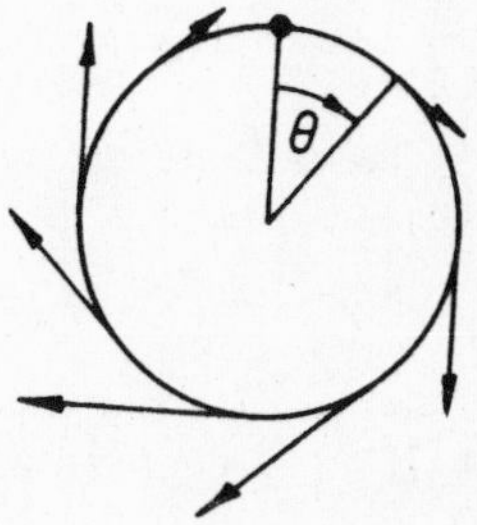

where $\theta \in [0,2\pi]$ is an angular coordinate for the circle such that $\theta=0$ and $\theta=2\pi$ represent the same point, which point is the only rest point of the differential equation. Since all solutions which start with small values of θ move away and pass through $\theta=\pi$, this rest point is *unstable*. However, *all* solutions approach $\theta=2\pi$ (i.e. the rest point) as $t \to \infty$.

3.1.3 Lyapunov functions

It is generally not possible to obtain the general solution of a non=linear system of differential equations, and hence to decide stability or otherwise of a rest point by actual inspection of solutions. The following two theorems of Lyapunov provide a direct method for proving stability without explicit knowledge of the general solution.

Let $\hat{x}$ be a rest point of the system

$$\dot{x} = f(x) \tag{3.1.3}$$

of differential equations.

DEFINITION 3.1.2 *Let* $V : \Omega \to \mathbb{R}$ *be a continuously differentiable scalar function on some neighbourhood* Ω *of* $\hat{x}$*.* V *is called a Lyapunov function* (for (3.1.3) *around* $\hat{x}$) *if*

(a) $V(\hat{x}) = 0$ *and* $V(x) > 0$ *for* $x \in \Omega \setminus \{\hat{x}\}$*; thus* V *has a strict minimum on* Ω *at* $\hat{x}$*; and*

(b) *the scalar function* $\dot{V} : \Omega \to \mathbb{R}$ *defined by*

$$\dot{V}(x) \triangleq \langle \nabla V(x), f(x) \rangle = \sum_{j=1}^{n} \frac{\partial V(x)}{\partial x^j} f_j(x)$$

satisfies the inequality

$$\dot{V}(x) \leqslant 0 \text{ for all } x \in \Omega$$

Note that if $x(\cdot)$ denotes any *solution* of (3.1.3), then

$$\dot{V}(x(t)) \equiv \frac{d}{dt} V(x(t)). \tag{3.1.4}$$

THEOREM 3.1.3 *If there exists a Lyapunov function for* (3.1.3) *around* $\hat{x}$*, then* $\hat{x}$ *is a stable rest point.*

PROOF Let $V : \Omega \to \mathbb{R}$ be a Lyapunov function around $\hat{x}$, and let U be any bounded neighbourhood of $\hat{x}$. Then we can find a (possibly smal= ler) neighbourhood W of $\hat{x}$ such that the closure $\bar{W}$ of W is contained in $U \cap \Omega$. Then the *boundary* ∂W of W is a compact set contained in $U \cap \Omega$, but not containing $\hat{x}$.

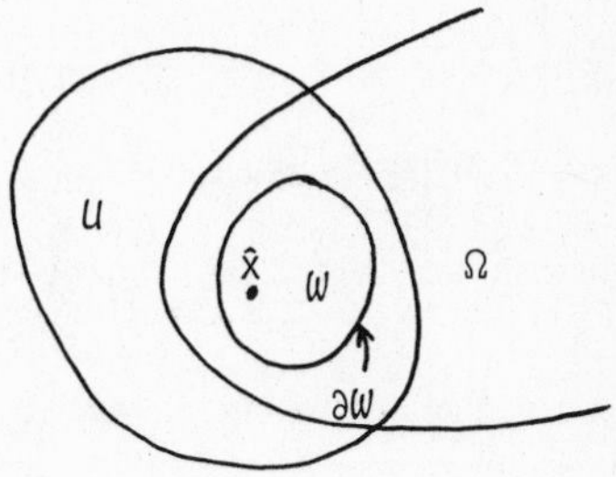

By (a), V is positive on ∂W, and, being continuous, its minimum value v on ∂W is positive. Thus there exists $v > 0$ such that

$$V(x) \geqslant v \text{ for all } x \in \partial W. \tag{3.1.5}$$

But since $V(\hat{x}) = 0$ and V is continuous, the set

$$V^{-1}[0,v) = \{x \in \Omega | V(x) < v\}$$

is a non-empty neighbourhood of $\hat{x}$, and hence so is

$$V = W \cap V^{-1}[0,v).$$

We show that owing to property (b), all solutions which start in V re= main in W, and hence in U. Let $x(\cdot)$ be a solution of (3.1.3), with $x(0) \in V$. This means that

$$x(0) \in W \text{ and } V(x(0)) < v. \tag{3.1.6}$$

It follows from (b) and (2.1.4) that

$$\frac{d}{dt} V(x(t)) \leqslant 0 \text{ for all } t \geqslant 0,$$

and hence that

$$V(x(t)) < v \text{ for all } t \geqslant 0.$$

Comparison with (3.1.5) shows that for no $t \geq 0$ can we have $x(t) \in \partial W$ It follows easily from (3.1.6) and the continuity of $x(\cdot)$ that we must have $x(t) \in W$ for all $t \geq 0$, as claimed. That $x(\cdot)$ is defined for all $t \geq 0$ follows from the fact that a solution which cannot leave a boun= ded set can always be extended.

The second theorem identifies an extra condition on the Lyapunov func= tion which is sufficient to guarantee asymptotic stability.

THEOREM 3.1.4 Suppose there exists a Lyapunov function $V: \Omega \to \mathbb{R}$ (3.1.3) *around* $\hat{x}$ *with the property that no solution* $x(\cdot)$ *of* (3.1.3) *lying in* Ω *(other than the rest solution* $x(t) \equiv \hat{x}$*) can satisfy*

$$V(x(t)) \equiv \text{constant, i.e. } \dot{V}(x(t)) \equiv 0. \qquad (3.1.7)$$

Then $\hat{x}$ *is an asymptotically stable rest point.*

PROOF Let U be any given bounded neighbourhood of $\hat{x}$. By the pre= vious theorem, there is a neighbourhood V of $\hat{x}$, with $V \subseteq U$, such that for any solution $x(\cdot)$ of (3.1.3),

$$x(0) \in V \text{ implies } x(t) \in U \text{ for all } t \geq 0.$$

We now show that

$$x(t) \to \hat{x} \text{ as } t \to \infty. \qquad (3.1.8)$$

Since $x(\cdot)$ is confined to the bounded set U, if (3.1.8) fails there must be an increasing sequence of times

$$t_k \uparrow \infty$$

and a point x_0 such that

$$x(t_k) \to x_0 \neq \hat{x}. \tag{3.1.9}$$

Since $V(x(\cdot))$ is non-increasing, it follows that for all $t \geqslant 0$

$$V(x(t)) \geqslant \lim_k V(x(t_k)) = V(x_0) > 0. \tag{3.1.10}$$

Consider the solution $\bar{x}(\cdot)$ starting at $\bar{x}(0) = x_0$, and suppose that at some time T, it satisfies

$$V(\bar{x}(T)) < V(x_0) = V(\bar{x}(0)). \tag{3.1.11}$$

By continuity of V, and the continuous dependence of solutions upon initial conditions, it follows that all solutions $\tilde{x}(\cdot)$ which start sufficiently close to $\bar{x}(0)$ will also satisfy

$$V(\tilde{x}(T)) < V(x_0).$$

In particular, by (3.1.9), for all sufficiently large k, this is true of the solution $\tilde{x}_k(\cdot)$ starting at $x(t_k)$. But of course

$$\tilde{x}_k(t) \equiv x(t_k+t), \quad t \geqslant 0$$

and thus for all sufficiently large k,

$$V(x(t_k+T)) < V(x_0).$$

This contradicts (3.1.10), thereby showing that (3.1.11) is impossible. It follows that the solution $\bar{x}(\cdot)$ satisfies

$$V(\bar{x}(\cdot)) \equiv V(x_0).$$

The extra hypothesis of the theorem immediately implies that $\bar{x}(\cdot)$ must be the rest solution, so that $x_0 = \hat{x}$. This contradicts (3.1.9), which finally shows that (3.1.8) must hold, as claimed.

Instability can also be established by this sort of direct method.

THEOREM 3.1.5 *Suppose* $\hat{x}$ *is a rest point for* (3.1.3). *Let* Ω *be a neighbourhood of* $\hat{x}$, $V: \Omega \to \mathbb{R}$ *a continuously differentiable function, and* $\Omega_1 \subseteq \Omega$ *an open set such that*

(a) $V(x) > 0$ and $\dot{V}(x) > 0$ in Ω_1;

(b) $V(x) = 0$ for $x \in \Omega \cap \partial\Omega_1$

(c) $\hat{x} \in \partial\Omega_1$

Under these conditions $\hat{x}$ *is an unstable rest point.*

We close this subsection with the remark that it has been established that, in principle, the question of the stability, asymptotic stability or instability of a rest point can always be decided using these methods.

3.1.4 Complete stability

A rest point $\hat{x}$ of the system (3.1.1) is said to be *completely stable* (or *globally asymptotically stable*) if it is asymptotically stable and *all* solutions of (3.1.1) tend to $\hat{x}$ as $t \to \infty$. Obviously $\hat{x}$ is then the only rest point.

It turns out that a globally defined Lyapunov function needs one further property to ensure complete stability. A real-valued function V on a topological space X is called LE-*compact* if for all $\alpha \in \mathbb{R}$, the lower-level set

$$\{x \in X \mid V(x) \leqslant \alpha\}$$

is compact. Of course, if X itself is compact then all continuous functions $V : X \to \mathbb{R}$ are LE-compact. On $\mathbb{R}^n$, a sufficient condition for a continuous function V to be LE-compact is that

$$V(x) \to \infty \text{ as } |x| \to \infty .$$

THEOREM 3.1.6 Let V *be a globally defined,* LE-*compact Lyapunov function for a rest point* $\hat{x}$ *of* (3.1.1) *with the property that no solution* $x(\cdot)$ *of* (3.1.1) *(other than the rest solution* $x(t) \equiv \hat{x}$*) can satisfy*

$$V(x(t)) \equiv \text{constant}.$$

Then $\hat{x}$ *is a completely stable rest point of* (3.1.1).

PROOF By Theorem 3.1.3, $\hat{x}$ is a stable rest point, and it remains to show that all solutions tend to $\hat{x}$. Let $x(\cdot)$ be any solution. By the basic properties of Lyapunov functions

$$V(x(t)) \leqslant V(x(0)) \text{ for all } t \geqslant 0,$$

which shows that $x(\cdot)$ is confined to the compact lower level set on which $V(x) \leqslant V(x(0))$. From this point on the proof is a copy of that of Theorem 3.1.4, with the role of the bounded set U being played by this compact (and hence bounded) lower level set.

3.1.5 Equilibria of conservative mechanical systems

Let $q = \hat{q}$ be a strict local minimum of the potential energy $V(\cdot)$ of a conservative mechanical system, with Hamiltonian

$$H(q,p) = T(q,\dot{q}) + V(q),$$

where the kinetic energy $T(q,\dot{q})$ is a positive definite, quadratic form in the generalized velocity vector $\dot{q}$, and the momenta

$$p \triangleq \partial T/\partial\dot{q}$$

are thus homogeneous linear expressions in $\dot{q}$.

The Hamiltonian equations of motion are

$$\dot{q} = \partial H/\partial p \quad , \quad \dot{p} = -\partial H/\partial q,$$

and $(q,p) = (\hat{q},0)$ is a rest point.

The function

$$H(q,p) - V(\hat{q})$$

serves as a Lyapunov function. Clearly, because $\hat{q}$ is a strict local minimizer for $V(\cdot)$, we have $H(q,p) > V(\hat{q})$ for all q is some neighbourhood of $\hat{q}$ and for all p, except $(q,p) = (\hat{q},0)$, while

$$\dot{H}(q,p) \triangleq \frac{\partial H}{\partial q}\frac{\partial H}{\partial p} + \frac{\partial H}{\partial p}(-\frac{\partial H}{\partial q}) \equiv 0.$$

Thus Theorem 3.1.3 establishes the stability of the rest point. Because H is constant along all solutions, $(\hat{q},0)$ is merely stable.

Further applications of these theorems are given in subsequent sec= tions.

3.2 LINEAR HOMOGENEOUS SYSTEMS

3.2.1 The eigenvalue criterion

The origin in $\mathbb{R}^n$ is obviously a rest point of any homogeneous linear system

$$\dot{x} = Ax, \tag{3.2.1}$$

where A is any real $n \times n$ matrix. The general solution of (3.2.1) is

$$x(t) = e^{At}x(0),$$

where e^{At} is the $n \times n$ matrix function defined by the convergent power series

$$e^{At} = I + At + \frac{1}{2!}A^2t^2 + \frac{1}{3!}A^3t^3 + \ldots = \sum_{k=0}^{\infty} \frac{1}{k!}A^kt^k.$$

It follows that the origin is stable iff $\|e^{At}\|$ is bounded as $t \to \infty$, and asymptotically stable iff $e^{At} \to 0$ as $t \to \infty$. As is well known, this matter is controlled by the location of the *eigenvalues* of A in the following way. The origin is:

asymptotically stable iff all eigenvalues of A have negative real part;

merely stable if there are simple (i.e. unrepeated) eigenvalues on the imaginary axis, the rest having negative real part;

unstable otherwise .

As noted in Lecture 1, if an $n \times n$ matrix A has all of its eigenvalues in the negative half-plane, it is called a *stability matrix* , this condition being equivalent to the condition that the origin be an asymptotically stable rest point of (3.2.1).

3.2.2 Quadratic Lyapunov functions for linear systems

Let us suppose that A in (3.2.1) is a stability matrix, and that we seek a Lyapunov function V. Since this requires

$$V(x) > 0 \text{ except when } x = 0,$$

one thinks naturally of trying a positive definite quadratic form

$$V(x) = x'Px, \tag{3.2.2}$$

with P symmetric positive definite. Then

$$\dot{V}(x) = \langle \nabla V(x), Ax \rangle = 2x'PAx = x'(PA + A'P)x.$$

Thus $\dot{V}$ is also a quadratic form, and we are led to the following ques= tion: given a stability matrix A, can one find a positive definite matrix P such that PA + A'P is negative definite? The equation

$$PA + A'P = - U \tag{3.2.3}$$

is the famous Lyapunov matrix equation, and the above question is ans= wered affirmatively by the following theorems.

THEOREM 3.2.1 *Suppose* A *is a stability matrix. Then, for any* U, *equation* (3.2.3) *has a unique solution for* P. *If* U *is positive definite, so is* P.

THEOREM 3.2.2 *Let* U = C'C *be positive semi-definite and such that the pair* (A,C) *is completely observable. Then if* A,P,U *satisfy* (3.2.3), A *is a stability matrix iff* P *is positive definite.*

PROOFS Suppose A is a stability matrix. The kernel of the l.h.s. of (3.2.3) is the set of all symmetric matrices P that satisfy

$$PA + A'P = 0.$$

Let P be any such matrix. Then since

$$\frac{d}{dt}(e^{A't} P \, e^{At}) = e^{A't}(P\,A + A'P)e^{At} = 0$$

we have

$$e^{A't}Pe^{At} \equiv P.$$

Since the left-hand side tends to zero as $t \to \infty$, it follows that P = 0. Thus (3.2.3), regarded as $\frac{1}{2}n(n+1)$ independent equations for $\frac{1}{2}n(n+1)$ independent entries of the symmetric matrix P, has zero kernel, and hence has a unique solution for any right-hand side U.

The unique solution of (3.2.3) is given explicitly by the integral

$$P = \int_0^\infty e^{A't}\, Ue^{At}\, dt. \qquad (3.2.4)$$

To show this, we evaluate

$$PA + A'P = \int_0^\infty e^{A't}(A'U + UA)e^{At}dt = \int_0^\infty \frac{d}{dt}(e^{A't}Ue^{At})dt$$

$$= [e^{A't}Ue^{At}]_0^\infty = - U.$$

Note that the infinite integral (3.2.4) exists because $e^{At} \to 0$ *exponentially* as $t \to \infty$.

It is also obvious from (3.2.4) that if U is positive definite, so is P, for then the integrand in (3.2.4) is positive definite.

Turning to Theorem 3.2.2, with $U = C'C$, and A a stability matrix, we have for any $x \in \mathbb{R}^n$

$$x'Px = \int_0^\infty x'e^{A't}C'Ce^{At}xdt = \int_0^\infty |Ce^{At}x|^2 dt \geqslant 0$$

with equality only if

$$Ce^{At}x = 0 \text{ for all } t \geqslant 0,$$

which holds only for $x = 0$ if (A,C) is completely observable. Thus in this case P, as given by (3.2.4), is positive definite. For the 'if' part, suppose P is positive definite, and let V be given by (3.2.2). Then, for any solution $x(t) = e^{At}x_0$ of (3.2.1), we have

$$\dot{V}(x(t)) = x_0'e^{A't}(A'P+PA)e^{At}x_0 = - |Ce^{At}x_0|^2 \leqslant 0,$$

which, again because of the complete observability assumption, cannot vanish identically unless $x_0 = 0$. Thus V is a Lyapunov function which satisfies the extra hypothesis of Theorem 3.1.4. Hence the origin is asymptotically stable, i.e. A is a stability matrix. This completes the proofs.

Theorem 3.2.2 will prove very useful in the sequel, while Theorem 3.2.1

solves the 'inverse Lyapunov problem' for linear homogeneous equations: if the origin is asymptotically stable for (3.2.1), then, choosing U to be any positive definite matrix, we solve (3.2.3) to find its unique solution P, which will be positive definite too. Then

$$V(x) = x'Px$$

is a quadratic Lyapunov function for which

$$\dot{V}(x) = -x'Ux.$$

An important consequence of this is given in the following subsection.

3.2.3 Asymptotic stability of nonlinear systems by linearization

We return now to the general nonlinear system (3.1.1) in $\mathbb{R}^n$, with a supposed rest point $\hat{x}$. Then

$$f(x) = A(x-\hat{x}) + g(x),$$

where A is the *Jacobian matrix* of the function $f:\mathbb{R}^n \to \mathbb{R}^n$ at the point $\hat{x}$, and

$$g(x) = o(|x-\hat{x}|) \text{ as } x \to \hat{x}. \qquad (3.2.5)$$

Thus if we 'linearized' the equations (3.1.1) about the rest point, we should obtain the linear system

$$\dot{z} = Az \qquad (3.2.6)$$

where we have written $z = x - \hat{x}$.

THEOREM 3.2.3 *If the Jacobian matrix* A *is a stability matrix, then* $\hat{x}$ *is an asymptotically stable rest point of* (3.1.1). *If* A *has one or more eigenvalues with positive real part,* $\hat{x}$ *is an unstable rest point of* (3.1.1).

PROOF We prove only the first part. The idea is to use Theorem 3.2.1 to construct a quadratic Lyapunov function for (3.2.6), and then to use the same function for (3.2.1).

Choosing U = I (the identity matrix) in (3.2.3), since A is a stability matrix, there is a unique positive definite P such that

$$PA + A'P = - I.$$

Turning to the nonlinear system (3.1.1), we let

$$V(x) = (x-\hat{x})'P(x-\hat{x}).$$

Then

$$\begin{aligned}\dot{V}(x) &= (x-\hat{x})'P(A(x-\hat{x}) + g(x)) + ((x-\hat{x})'A' + g'(x))P(x-\hat{x}) \\ &= (x-\hat{x})'(PA+A'P)(x-\hat{x}) + 2(x-\hat{x})'Pg(x) \\ &= -|x-\hat{x}|^2 + 2(x-\hat{x})'Pg(x).\end{aligned}$$

Because of (3.2.5), the first term here is dominant as $x \to \hat{x}$. More precisely,

$$|2(x-\hat{x})'Pg(x)| \leqslant 2|x-\hat{x}|\|P\|\,|g(x)|,$$

and, by (3.2.5), there exists $r > 0$ such that

$$|g(x)| \leqslant |x-\hat{x}|/4\|P\| \text{ whenever } |x-\hat{x}| < r,$$

so that

$$|2(x-\hat{x})'Pg(x)| \leqslant \tfrac{1}{2}|x-\hat{x}|^2.$$

Thus taking the region Ω in Definition 3.1.2 as the ball $|x-\hat{x}| < r$, we see that $V(x) > 0$ on $\Omega \setminus \{\hat{x}\}$ and

$$\dot{V}(x) \leqslant -\tfrac{1}{2}\,|x-\hat{x}|^2.$$

Thus $V:\Omega \to \mathbb{R}$ is a Lyapunov function for (3.1.1) around $\hat{x}$, and further=more, the extra hypothesis of Theorem 3.1.4 is satisfied. Consequently, by that theorem, $\hat{x}$ is an asymptotically stable rest point for (3.1.1), as claimed.

REMARKS

1. No conclusions regarding $\hat{x}$ can be drawn if (3.2.6) turns out to be merely stable.

2. For systems (3.1.1) on a general differentiable manifold the same result holds. At a rest point $\hat{x}$ of the vector field f the Jacobian matrix $(\partial f^i(\hat{x})/\partial x^j)$ in any local coordinate system defines a second-rank mixed tensor, the eigenvalues of which are invariantly associated with the rest point. If all eigenvalues have negative real parts, the above proof can be carried through in any local coordinate system.

3.3 ABSOLUTE STABILITY OF CONTROL SYSTEMS

The term absolute stability refers to the ability of an already asymp=totically stable linear system to remain asymptotically stable despite the introduction of an *unknown* dynamic nonlinearity of a certain type.

We consider a single-input, single-output linear control system in $\mathbb{R}^n$

$$\begin{aligned} \dot{x} &= Ax + bu \\ y &= c'x \quad , \end{aligned} \tag{3.3.1}$$

where A is a stability matrix, and b,c are n-vectors.

To improve the stability, or for other reasons (eg. excessive overshoot or rise time) a regulator is to be used comprising a servo-motor or like device, often with a nonlinear characteristic, together with an integrator, to generate a control signal u from the measured scalar output y. A block diagram of such a regulator appears below.

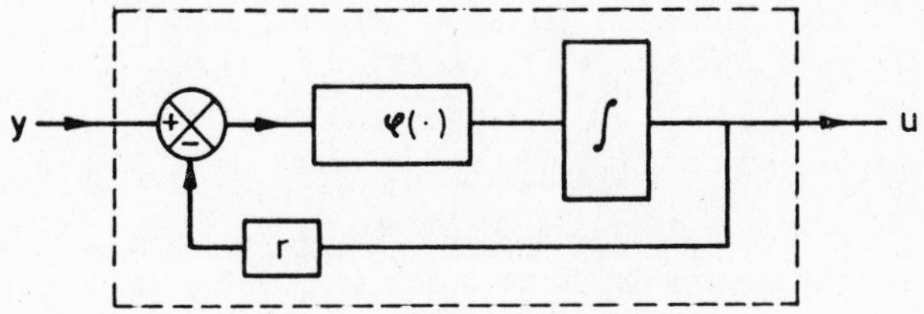

Here $\varphi(\cdot)$ is the nonlinear characteristic of the servo device, while r is a scalar parameter which we can choose.

Clearly

$$u(t) = \int_0^t \varphi(y - ru(\tau))d\tau,$$

or

$$\dot{u} = \varphi(y - ru).$$

Thus the automatically controlled system has state dimension n + 1, with state variables (x,u) and equations

$$\begin{aligned} \dot{x} &= Ax + bu \\ \dot{u} &= \varphi(y - ru). \end{aligned} \tag{3.3.2}$$

The nonlinear characteristic $\varphi(\cdot)$ may be only very imperfectly known, or subject to unpredictable variations, but usually has the following properties, which we assume:

(i) $\varphi(0) = 0$ (no signal - no control);

(ii) $\varphi(\sigma)$ has the same sign as σ;

(iii) $\int_0^{\pm\infty} \varphi(\sigma)d\sigma$ diverge to ∞;

(iv) φ is continuous.

DEFINITION 3.3.1 *The quadruple* (A,b,c,r) *is absolutely stable if the system* (3.3.2) *is completely stable for all choices of nonlinear characteristic* φ *having properties* (i) - (iv) .

We close this lecture by using Lyapunov's theorems to give sufficient conditions for absolute stability. First, we select any positive de= finite n×n matrix U, and let P be the unique solution of the Lyapunov matrix equation

$$A'P + PA = - U.$$

By Theorem 3.2.1, since A is a stability matrix, P is unique and posi= tive definite. Next we let

$$\Phi(\sigma) \triangleq \int_0^{\sigma} \varphi(\tau)d\tau.$$

Because of (i) and (ii) we have

$$\Phi(\sigma) > 0 \text{ unless } \sigma = 0, \quad \Phi(0) = 0. \tag{3.3.3}$$

Now we define

$$V(x,u) = (Ax+bu)'P(Ax+bu) + \Phi(c'x-ru). \tag{3.3.4}$$

Since P is positive definite, it is clear from (3.3.3) that

$$V(x,u) > 0 \text{ unless } Ax + bu = 0 \text{ and } c'x-ru=0.$$

Provided the determinant

$$\begin{vmatrix} A & b \\ c' & -r \end{vmatrix} \neq 0 \tag{3.3.5}$$

as we henceforth assume, it follows that

$$V(x,u) > 0 \text{ unless } (x,u) = (0,0),$$

while obviously, $V(0,0) = 0$. Furthermore because of condition (iii) and (3.3.5), we have

$$V(x,u) \to \infty \quad \text{as} \quad |x| + |u| \to \infty,$$

so that V is LE-compact on $\mathbb{R}^{n+1}$. Finally we consider

$$\dot{V}(x,u) = (A\dot{x} + b\dot{u})'P(Ax+bu) + (Ax+bu)'P(A\dot{x}+b\dot{u})$$

$$+ \varphi(c'x-ru)(c'\dot{x}-r\dot{u}).$$

Writing

$$z = Ax + bu \text{ and } \sigma = c'x-ru, \tag{3.3.6}$$

and using (3.3.2), this simplifies to

$$\dot{V}(x,u) = (Az + b\varphi(\sigma))'Pz + z'P(Az + b\varphi(\sigma))$$

$$+ \varphi(\sigma)(c'z - r\varphi(\sigma))$$

$$= -z'Uz + \varphi(\sigma)b'Pz + z'Pb\varphi(\sigma) + \varphi(\sigma)c'z - r\varphi^2(\sigma),$$

or, in matrix form,

$$\dot{V}(x,u) = - [z',\varphi(\sigma)] \left[\begin{array}{c|c} U & -Pb-\frac{1}{2}c \\ \hline -b'P-\frac{1}{2}c' & r \end{array}\right] \begin{bmatrix} z \\ \varphi(\sigma) \end{bmatrix}. \tag{3.3.7}$$

Thus $\dot{V}$ is a quadratic form in z, $\varphi(\sigma)$. Since U is positive definite, the coefficient matrix in (3.3.7) is positive definite iff the (n+1) x (n+1) determinant

$$\left|\begin{array}{c|c} U & -Pb-\frac{1}{2}c \\ \hline -b'P-\frac{1}{2}c' & r \end{array}\right| > 0. \qquad (3.3.8)$$

Assuming this, it then follows from (3.3.7), (3.3.6) and (3.3.5) that

$$\dot{V}(x,u) < 0 \text{ unless } (x,u) = (0,0).$$

But then all the conditions of Theorem 3.1.6 are satisfied, and the origin $(x,u) = (0,0)$ is a completely stable rest point for (3.3.2). Since this holds for any characteristic $\varphi(\cdot)$ satisfying (i) - (iv), this proves absolute stability for any quadruple (A,b,c,r) for which there exist positive definite matrices U,P satisfying the Lyapunov matrix equation and (3.3.5) and (3.3.8).

The latter conditions can be simplified. Since A is a stability matrix it is invertible, and we may use the identity

$$\left[\begin{array}{c|c} A & b \\ \hline c' & -r \end{array}\right]\left[\begin{array}{c|c} A^{-1} & -A^{-1}b \\ \hline 0 & 1 \end{array}\right] = \left[\begin{array}{c|c} I & 0 \\ \hline c'A^{-1} & -c'A^{-1}b-r \end{array}\right]$$

to deduce, by taking determinants, that (3.3.5) is equivalent to

$$r + c'A^{-1}b \neq 0. \qquad (3.3.9)$$

Similarly, since U has positive determinant , (3.3.8) is equivalent to

$$r > (Pb + \tfrac{1}{2}c)'U^{-1}(Pb + \tfrac{1}{2}c). \qquad (3.3.10)$$

We summarize in our final theorem.

THEOREM 3.3.2 *Assuming that we are given the system* (2.3.1) *with* A *a stability matrix, and have chosen any positive definite matrix* U *, let* P *be the unique solution of*

$$A'P + PA = -U. \tag{3.3.11}$$

Then for any choice of the regulator feedback gain r *satisfying* (3.3.9) *and* (3.3.10), *the quadruple* (A,b,c,r) *will be absolutely stable.*

This raises the question of the infimum of the expression

$$(Pb + \tfrac{1}{2}c)'U^{-1}(Pb+\tfrac{1}{2}c),$$

subject to (3.3.11), over all positive definite matrices U. Lefshetz has shown that this infimum is equal to

$$c'b + \sup_{-\infty < \omega < \infty} \operatorname{Im}(\omega c'(i\omega I - A)^{-1}b).$$

3.4 BIBLIOGRAPHY

The sources from which this lecture was prepared are:

J.P. La Salle and S. Lefschetz, *Stability by Liapunov's Direct Method.* Academic Press, 1961.

S. Lefschetz, *Stability of Nonlinear Control Systems.* Academic Press, 1965.

The reader should consult them for further developments and ramifications More recent books include:

W. Hahn, *Stability of Motion.* Springer-Verlag, New York, 1967.

J.P. La Salle, *The Stability of Dynamical Systems.* SIAM Publications, Philadelphia, 1976.

V.I. Arnold, *Mathematical Methods of Classical Mechanics.* Springer-Verlag, New York, 1978.

4. LINEAR-QUADRATIC OPTIMAL CONTROL

4.1 INTRODUCTION

In this lecture we present an account of the beautiful theory, due in essence to R.E. Kalman, of the use of a quadratic *performance criterion* to guide the selection of a feedback law for a linear control system. Because the case of optimal stabilization of a time-invariant system is by far the most important, we give a self-contained treatment of it which includes relevant computational techniques. This discussion is followed by more concise sections on time-varying and discrete-time problems.

4.2 THE TIME-INVARIANT LINEAR-QUADRATIC PROBLEM

4.2.1 Quadratic performance criteria

Let

$$\dot{x}(t) = Ax(t) + Bu(t) \tag{4.2.1}$$

be a stabilizable time-invariant linear system, in which $x = 0$ repre= sents the desired (zero-perturbation) state, and suppose we wish to choose a feedback law

$$u(t) = -Kx(t)$$

which not only asymptotically stabilizes (4.2.1), but is a 'good' choice among all such feedback laws. As a measure of the efficacy of a parti= cular control function

$$u(t),\ t \geqslant o$$

in restoring a disturbed (i.e. non-zero) initial state x_o, we might adopt a performance criterion of the form

$$V(x_0,u(\cdot)) = \int_0^\infty [u'(t)Ru(t) + x'(t)Qx(t)]\, dt, \qquad (4.2.2)$$

in which of course $x(t)$, $t \geq 0$ denotes the response of the system (4.2.1) to the controller $u(\cdot)$, from the initial state x_0.

In (4.2.2), R should be a *positive definite* symmetric $r \times r$ matrix, the term $u'(t)Ru(t)$ serving to penalize unduly heavy control action, while Q may be a *positive semidefinite* symmetric $n \times n$ matrix, the term $x'(t)Qx(t)$ serving to penalize large or persistent undesirable devia= tions in $x(t)$. Thus a 'good' controller for dealing with the deviation x_0 is one with a low value of $V(x_0,u(\cdot))$, and we should like to choose $u(\cdot)$ as that controller, if it exists, which minimizes $V(x_0,u(\cdot))$ in the class of all controllers $u(\cdot)$ for which

$$x(t) \to 0 \text{ as } t \to \infty. \qquad (4.2.3)$$

4.2.2 Solution of the optimal control problem

To solve this optimal control problem we use a device analogous to square-completion, which goes back to Legendre's studies in the calculus of variations. Let P be any real $n \times n$ symmetric matrix. Then with $x_0, u(\cdot)$ and $x(\cdot)$ as above, we have

$$\frac{d}{dt}[x'(t)Px(t)] = (Ax(t)+Bu(t))'Px(t) + x'(t)P(Ax(t)+Bu(t)). \quad (4.2.4)$$

To save space we shall no longer explicitly show the t-dependence of $x(\cdot)$ and $u(\cdot)$. Integrating (4.2.4) from $t = 0$ to $t = T$ we obtain, since $x(0) = x_0$:

$$x'(T)Px(T) - x_0'Px_0 = \int_0^T [x'(A'P+PA)x + 2u'B'Px]\, dt.$$

Since we consider only controllers for which (4.2.3) holds (and such exist because (4.2.1) is supposed stabilizable) this shows that

$$\int_0^\infty [x'(A'P+PA)x + 2u'B'Px]\,dt = -x_0'Px_0.$$

Adding this to (4.2.2) we obtain for any admissible[1] controller

$$V(x_0,u(\cdot)) = \int_0^\infty [u'Ru+2u'B'Px + x'(A'P+PA+Q)x]\,dt + x_0'Px_0$$

or

$$V(x_0,u(\cdot)) = \int_0^\infty [(u+R^{-1}B'Px)'R(u+R^{-1}B'Px)+x'(A'P+PA-PBR^{-1}B'P+Q)x]\,dt$$
$$+ x_0'Px_0. \qquad (4.2.5)$$

Thus far the symmetric matrix P, which has been used to complicate things in this way, is entirely arbitrary. However *suppose* that P may be chosen to satisfy the quadratic matrix equation

$$A'P + PA - PBR^{-1}B'P + Q = 0, \qquad (4.2.6)$$

known as the *algebraic Riccati equation*. For this choice of P the performance index reduces to

$$V(x_0,u(\cdot)) = \int_0^\infty (u+R^{-1}B'Px)'R(u+R^{-1}B'Px)\,dt + x_0'Px_0.$$

Since R is positive definite, the integrand here is non-negative, and would be zero iff we could choose u(•) such that it and its response x(•) satisfy

$$u(t) + R^{-1}B'Px(t) \equiv 0, \quad t \geqslant 0. \qquad (4.2.7)$$

This would mean that x(•) satisfies the homogeneous system

$$\dot{x}(t) \equiv (A - BR^{-1}B'P)x(t)$$

---

1) i.e. one for which (4.2.3) holds.

or

$$x(t) = e^{\bar{A}t}x_0 \qquad (4.2.8)$$

where we have written

$$\bar{A} = A - BR^{-1}B'P. \qquad (4.2.9)$$

Since we require (4.2.3), regardless of x_0, it follows from (4.2.8) that $\bar{A}$ will have to be a stability matrix. Thus to summarize: *provided the algebraic Riccati equation* (4.2.6) *has a symmetric solution* P *for which* $\bar{A}$ *given by* (4.2.9) *is a stability matrix*, then by (4.2.7) and (4.2.8) the optimal controller for given x_0 is given in terms of this solution P by

$$u_*(t) = -R^{-1}B'Pe^{\bar{A}t}x_0$$

while the minimum value of the performance criterion is

$$V_*(x_0) = x_0'Px_0. \qquad (4.2.10)$$

Furthermore, since the optimal controller and its response satisfy (4.2.7), the optimal controller will be automatically generated by the feedback law

$$\left.\begin{aligned} u &= -Kx \\ \text{with} \qquad K &= R^{-1}B'P. \end{aligned}\right\} \qquad (4.2.11)$$

4.2.3 The condition on Q

Being positive semidefinite, the matrix Q may be factored in the form

$$Q = C'C,$$

and then the performance index may be written as

$$V(x_0,u(\cdot)) = \int_0^\infty [u'(t)Ru(t) + |y(t)|^2]dt, \qquad (4.2.12)$$

where we have put

$$y(t) = Cx(t). \qquad (4.2.13)$$

In the following subsections we shall justify the solution given above by showing that the algebraic Riccati equation does indeed have a (unique) solution P for which $\bar{A}$ given by (4.2.9) is a stability matrix, *under the extra condition on* Q *that the system* (4.2.1) *with output given by* (4.2.13) *is completely observable* . As noted in Lecture 1, this is equivalent to the condition

$$\text{rank } [C' \,\vdots\, A'C' \,\vdots\, \text{-----} \,\vdots\, A'^{n-1}C'] \quad = n. \qquad (4.2.14)$$

This condition is assumed for the remainder of Section 4.2. Note that it is certainly satisfied if Q is positive definite.

REMARK. Suppose a given controller $u(\cdot)$ has a *finite* performance index $V(x_0,u(\cdot))$. A standard necessary condition for convergence of the infinite integral (4.2.12) is that the integrand tends to zero. Since R is positive definite it follows that

$$u(t) \to 0 \text{ and } y(t) \to 0 \text{ as } t \to \infty.$$

However it can be shown that if (4.2.14) holds, then the finiteness of $V(x_0,u(\cdot))$ implies in addition that

$$x(t) \to 0 \text{ as } t \to \infty.$$

4.2.4 A crucial lemma

Let A,B,C,R *be matrices as above, with* R *positive definite and the pair* (A,C) *being completely observable* (*i.e. satisfying*(4.2.14)), *and suppose*

K,P,S *are matrices such that*

(a) K is $r\times n$; P is $n\times n$ *symmetric*; S is $n\times n$ *symmetric positive semidefinite*;

(b) *the equation*

$$(A-BK)'P + P(A-BK) = -K'RK - C'C - S \tag{4.2.15}$$

holds. Then $A-BK$ *is a stability matrix iff* P *is positive definite, and in this case* P *is unique as a solution of* (4.2.15).

The proof of this lemma is an interesting but not difficult exercise in the application of Theorem 3.2.2, and is left to the reader.

4.2.5 Properties of solutions of the algebraic Riccati equation

THEOREM 4.2.1 Let P *be any solution of* (4.2.6) *and set* $K = R^{-1}B'P$. *Then* $A-BK$ *is a stability matrix iff* P *is positive definite.*

PROOF Using (4.2.6) we have

$$\begin{aligned}(A-BK)'P + P(A-BK) &= A'P + PA - 2PBR^{-1}B'P\\ &= -PBR^{-1}B'P - Q\\ &= -K'RK - C'C.\end{aligned}$$

This is of the form (4.2.15) with $S = 0$, and the result follows immediately from the lemma. ∎

THEOREM 4.2.2 There cannot be more than one positive definite solution of (4.2.6)

PROOF If P_1 and P_2 were both positive definite symmetric solutions of (4.2.6), then, by the lemma, both would yield stability matrices (4.2.9), and hence by (4.2.10), for any x_o we should have

$$x_o'P_1x_o = V_*(x_o) = x_o'P_2x_o,$$

whence $P_1 = P_2$. ∎

THEOREM 4.2.3 Any solution P *of* (4.2.6) *is invertible.*

PROOF Suppose $P\xi = 0$ for some $\xi \in \mathbb{R}^n$. Then multiplying (4.2.6) on the left by ξ' and on the right by ξ we obtain $\xi'C'C\xi = 0$, which implies $C\xi = 0$. Hence multiplying (4.2.6) on the right only by ξ, we now find $PA\xi = 0$. Repeating the argument with ξ replaced by $A\xi$, it follows that also $CA\xi = 0$ and $PA^2\xi = 0$. Continuing thus we prove that $P\xi = 0$ implies $C\xi = 0$, $CA\xi = 0, \ldots, CA^{n-1}\xi = 0$, or

$$\xi'[C' \,\vdots\, A'C' \,\vdots\, \text{----} \,\vdots\, A'^{n-1}C'] = 0.$$

By (4.2.14), this implies $\xi = 0$, showing that P is invertible. ∎

4.2.6 Kleinman's iteration

We turn now to the proof that condition (4.2.14) guarantees that there is a positive definite solution to the algebraic Riccati equation. There are several ways of doing this, and we base our proof on an iterative method for actually solving (4.2.6), due to D.L. Kleinman.

As in the proof of Theorem 3.2.1, if P is a solution of (4.2.6) and we set

$$K = R^{-1}B'P, \tag{4.2.16}$$

then (4.2.6) can be written in the form

$$(A-BK)'P + P(A-BK) = -K'RK - C'C. \tag{4.2.17}$$

In other words, P satisfies (4.2.6) iff P and K together satisfy (4.2.16) and (4.2.17). In Kleinman's method, we select a first guess at K, solve the *linear* matrix equation (4.2.17) for P, and then use this P to get a revised guess at K from (4.2.16), and repeat, alternately getting a new K and then a new P.

Step 0 Select any matrix K_1 for which $A-BK_1$ is a stability matrix. This is possible, since the system (4.2.1) is stabilizable.

Step i Having matrix K_i for which $A-BK_i$ is a stability matrix, solve the (Liapunov type) linear matrix equation

$$(A-BK_i)'P_i + P_i(A-BK_i) = -K_i'RK_i - C'C \tag{4.2.18}$$

for the symmetric matrix P_i, and then set

$$K_{i+1} = R^{-1}B'P_i. \tag{4.2.19}$$

This generates two sequences of matrices: $K_1K_2,\ldots;\ P_1,P_2,\ldots$

THEOREM 4.2.4 The iteration is well-defined, in that at each step the equation (4.2.18) *has a unique solution* P_i, *which is positive de= finite, and the matrix* K_{i+1} *given by* (4.2.19) *is then such that* $A-BK_{i+1}$ *is a stability matrix.*

PROOF Suppose that at the beginning of step i we have K_i such that $A-BK_i$ is a stability matrix. (This is true for i = 1.) By the lemma (with S = 0)(4.2.18) has a unique solution P_i, and it is positive definite. Then by (4.2.18) and (4.2.19) we have

$$\begin{aligned}(A-BK_{i+1})'P_i + P_i(A-BK_{i+1}) &= A'P_i + P_iA - K_{i+1}'B'P_i - P_iB'K_{i+1}\\ &= -K_i'RK_i - C'C - (K_{i+1}-K_i)'B'P_i - P_iB(K_{i+1}-K_i)\\ &= -K_i'RK_i - C'C - (K_{i+1}-K_i)'RK_{i+1} - K_{i+1}'R(K_{i+1}-K_i)\\ &= -K_{i+1}'RK_{i+1} - C'C - (K_{i+1}-K_i)'R(K_{i+1}-K_i).\end{aligned} \tag{4.2.20}$$

The final equation here is of the form (4.2.15) with

$$S = (K_{i+1}-K_i)'R(K_{i+1}-K_i),$$

and hence by the lemma, since P_i is positive definite, $A-BK_{i+1}$ is a stability matrix. By induction, all P_i are uniquely determined and po=sitive definite, and all $A-BK_i$ are stability matrices. ■

Our final theorem of this section establishes the convergence of the algorithm, and hence the existence of a positive definite solution of the algebraic Riccati equation.

THEOREM 4.2.5 *For any initial matrix* K_1 *as in Step 0, the sequences* K_i, P_i *converge:*

$$K_i \to K, \; P_i \to P,$$

where P *is a positive definite solution of the algebraic Riccati equation* (4.2.6) *and* A-BK *is a stability matrix.*

PROOF If we rewrite (4.2.18) with i replaced by i+1, multiply the result by λ and subtract from (4.2.20), we obtain

$$(A-BK_{i+1})'(P_i-\lambda P_{i+1}) + (P_i-\lambda P_{i+1})(A-BK_{i+1})$$

$$= -K'_{i+1}\tilde{R}K_{i+1} - \tilde{C}'\tilde{C} - (K_{i+1}-K_i)'R(K_{i+1}-K_i)$$

where $\tilde{R} = (1-\lambda)R$ and $\tilde{C} = \sqrt{1-\lambda}\,C$. This relationship is of the form (4.2.15), and for any $\lambda < 1$ the modified matrices $\tilde{R}$, $\tilde{C}$ still satisfy the provisos of the lemma. Since $A-BK_{i+1}$ is a stability matrix, it follows from the lemma that

$$P_i - \lambda P_{i+1}$$

is a positive definite matrix for all $\lambda < 1$, and hence that $P_i - P_{i+1}$ is positive semidefinite. Thus for every fixed $x \in \mathbb{R}^n$ the positive scalar sequence $x'P_ix$ is monotone decreasing, and hence tends to a limit. It is an easy consequence of this that the matrix sequence P_i must tend to a limit, P say, necessarily positive semidefinite. Then from (4.2.19) it follows that

$$K_i \to K = R^{-1}B'P,$$

and hence from (4.2.18) that P and K also satisfy (4.2.17). Hence P is a solution of the algebraic Riccati equation, and so being positive semidefinite and invertible (Theorem 4.2.3), P must be positive definite. Hence finally by Theorem 4.2.1, A-BK is a stability matrix. ∎

REMARKS Kleinman's iteration is equivalent to Newton's method applied to (4.2.6), so that the convergence is in fact quadratic. The above convergence theorem is thus a remarkable example of a *semi-global* convergence theorem for a Newton iteration. The method is often used in practice. If the matrix A is itself a stability matrix, one may at Step 0 choose $K_1 = 0$.

There are other methods of solving (4.2.6) - see the Historical Notes and References at the end of this lecture.

4.2.7 Summary for time-invariant problems

Having selected a performance criterion (4.2.2) for which the observability condition (4.2.14) holds (where Q = C'C), the algebraic Riccati equation (4.2.6) has a unique positive definite solution P, which may be found by Kleinman's iteration. The optimal control is then given in feedback form by

$$u = -Kx \text{ with } K = R^{-1}B'P,$$

while the minimum of V, for given initial state x_0, is

$$V_*(x_0) = x_0'Px_0.$$

The closed-loop system matrix

$$\bar{A} = A - BK$$

is a stability matrix.

4.2.8 Guaranteed stability margin

The beauty of the above result is that almost any choice of performance index will result in an asymptotically stable closed loop. Indeed, if (4.2.1) is completely controllable, a simple modification of this pro= cedure can guarantee that every closed-loop pole will have its real part less than a pre-assigned number $-\alpha$. To achieve this, we solve the modified Riccati equation

$$(A+\alpha I)'P + P(A+\alpha I) - PBR^{-1}B'P + C'C = 0 \qquad (4.2.21)$$

instead of (4.2.6). It is easily checked that if (4.2.1) is completely controllable, then so is the system

$$\dot{x} = (A+\alpha I)x + Bu \qquad (4.2.22)$$

and similarly for complete observability with output (4.2.13). Hence (4.2.21) has a unique positive definite solution P_α, and the feedback law

$$u = K_\alpha x = R^{-1}B'P_\alpha x \qquad (4.2.23)$$

renders the closed-loop system matrix of (4.2.22) i.e.

$$\bar{A}_\alpha = A + \alpha I - B'R^{-1}B'P_\alpha$$

a stability matrix. Since the term αI here simply adds α to each eigen= value, it follows that the closed-loop system matrix

$$A - B'R^{-1}B'P_\alpha$$

of the actual system (4.2.1) with control law (4.2.23), has eigenvalues with real part less than $-\alpha$, as asserted. Of course one must expect that the entries for matrix K_α will become larger the larger we take α - more drastic stabilization will require heavier control action.

4.2.9 Example

Let us take

$$\begin{bmatrix} \dot{x}_1 \\ \dot{x}_2 \end{bmatrix} = \begin{bmatrix} 1 & 0 \\ 2 & 0 \end{bmatrix} \begin{bmatrix} x_1 \\ x_2 \end{bmatrix} + \begin{bmatrix} 1 \\ 0 \end{bmatrix} u \qquad (4.2.24)$$

and

$$V = \int_0^\infty (u^2 + 15x_1^2 + 25x_2^2)dt.$$

Here $Q = \begin{bmatrix} 15 & 0 \\ 0 & 25 \end{bmatrix}$ is positive definite, so the observability condition (4.2.14) is satisfied. Also

$$\text{rank } [B \,\vdots\, AB] = \text{rank} \begin{bmatrix} 1 & 1 \\ 0 & 2 \end{bmatrix} = 2,$$

so (4.2.24) is completely controllable, and hence certainly stabilizable. The eigenvalues of $A = \begin{bmatrix} 1 & 0 \\ 2 & 0 \end{bmatrix}$ are 0 and 1 - A is not a stability matrix. To find a stabilizing feedback with which to start Kleinman's iteration, we consider

$$A - BK = \begin{bmatrix} 1 & 0 \\ 2 & 0 \end{bmatrix} - \begin{bmatrix} 1 \\ 0 \end{bmatrix} [k_1 \quad k_2] = \begin{bmatrix} 1-k_1 & -k_2 \\ 2 & 0 \end{bmatrix}.$$

The characteristic polynomial is

$$\lambda^2 + (k_1-1)\lambda + 2k_2$$

which reduces, for example, to $(\lambda+2)^2$ for $k_1=5$ and $k_2=2$. Thus $K = [5,2]$ is a stabilizing gain matrix.

The iteration equations (4.2.18) and (4.2.19) for this case are

$$\begin{bmatrix} 1-k_1 & 2 \\ -k_2 & 0 \end{bmatrix}\begin{bmatrix} P_{11} & P_{12} \\ P_{12} & P_{22} \end{bmatrix} + \begin{bmatrix} P_{11} & P_{12} \\ P_{12} & P_{22} \end{bmatrix}\begin{bmatrix} 1-k_1 & -k_2 \\ 2 & 0 \end{bmatrix} = \begin{bmatrix} -15-k_1^2 & -k_1k_2 \\ -k_1k_2 & -25-k_2^2 \end{bmatrix}$$

$$[k_1,k_2] = (1)^{-1}\,[1,0]\begin{bmatrix} P_{11} & P_{12} \\ P_{12} & P_{22} \end{bmatrix} = [P_{11},P_{12}] \,. \qquad (4.2.25)$$

Here the first equation gives three independent equations for the three independent entries of the matrix P_i at each iteration: the equations are

$$\left.\begin{aligned} 2(1-k_1)P_{11} + 4P_{12} \qquad\qquad &= -15 - k_1^2 \\ -k_2P_{11} + (1-k_1)P_{12} + 2P_{22} &= \quad -k_1k_2 \\ -\,2k_2P_{12} \qquad\qquad &= -25 - k_2^2 \end{aligned}\right\} .$$

Thus at each iteration we start with a given (k_1,k_2), solve these equa= tions for P_{11},P_{12},P_{22}, update $[k_1,k_2]$ by (4.2.25), and then repeat. The following table shows the results, commencing at step 0 with $K = [5,2]$ as selected above.

Iteration Number	P_{11}	P_{12}	P_{22}
1	8.62500	7.25000	18.12500
2	7.26473	5.34914	15.46262
3	7.00923	5.01139	15.01417
4	7.00001	5.00001	15.00002
5	7.00000	5.00000	15.00000

It is easily checked that, as is suggested by these results, the posi= tive definite solution of the algebraic Riccati equation is

$$P = \begin{bmatrix} 7 & 5 \\ 5 & 15 \end{bmatrix} ,$$

giving the optimal gain matrix

$$K = [7,5] \,.$$

This results in the closed-loop system matrix

$$\bar{A} = A - BK = \begin{bmatrix} -6 & -5 \\ 2 & 0 \end{bmatrix},$$

the eigenvalues of which are the complex conjugate pair (with negative real part, as expected) $-3 \pm i$.

Finally we remark that the algebraic Riccati equation for this example has one other solution, viz the negative definite matrix

$$\begin{bmatrix} -5 & 5 \\ 5 & -15 \end{bmatrix}.$$

4.3 THE GENERAL LINEAR-QUADRATIC PROBLEM

4.3.1 Description of the problem

Consider the general time-varying linear system

$$\dot{x}(t) = A(t)x(t) + B(t)u(t). \qquad (4.3.1)$$

The matrix functions $A(\cdot)$ and $B(\cdot)$ are supposed to have bounded measurable entries - in most applications they are piecewise continuous - and the final time T is given. If at some time $t_0 \in [0,T)$ the state x_0 is different from the desired zero state, we might use a performance criterion of the form

$$V(t_0,x_0,u(\cdot)) = \int_{t_0}^{T} [u'(t)R(t)u(t) + x'(t)Q(t)x(t)]\,dt + x'(T)Q_f x(T) \qquad (4.3.2)$$

to measure the efficacy of any controller $u(t)$, $t_0 \leqslant t \leqslant T$ in improving matters. Here $R(\cdot)$ and $Q(\cdot)$ are chosen continuous symmetric matrix functions on $[0,T]$, with $R(t)$ *positive definite* and $Q(t)$ *positive semidefinite* for all t. The matrix Q_f should also be symmetric and positive semidefinite - the rôle of this term is to penalize any re=

maining state deviation at the final time. As before, a 'good' con=troller is one with a low value of V, and an optimal controller for given (t_0,x_0) is one which minimizes V.

4.3.2 Solution of the optimal control problem

The procedure is the same as for the previous case in Section 4.2.2. For *any* continuous and piecewise continuously differentiable symmetric n×n matrix function P(t), $0 \leqslant t \leqslant T$, we 'complicate' $V(t_0,x_0,u(\cdot))$ just as before, obtaining (c.f. (4.2.5))

$$V(t_0,x_0,u(\cdot)) = x_0'P(t_0)x_0 + x'(T)(Q_f-P(T))x(T) +$$
$$\int_{t_0}^{T} [(u+R^{-1}B'Px)'R(u+R^{-1}B'Px) + x'(\dot{P}+A'P+PA-PBR^{-1}B'P+Q)x]\,dt,$$

which reduces to

$$V(t_0,x_0,u(\cdot)) = \int_{t_0}^{T} (u+R^{-1}B'Px)'R(u+R^{-1}B'Px)dt + x_0'P(t_0)x_0 \qquad (4.3.3)$$

provided that $P(\cdot)$ may be chosen to be a solution of the Cauchy problem

$$\dot{P} + A'(t)P + PA(t) - PB(t)R^{-1}(t)B(t)P + Q(t) = 0 \qquad (4.3.4)$$

$$0 \leqslant t < T$$

$$P(T) = Q_f. \qquad (4.3.5)$$

Proceeding on this supposition (which we justify in the next subsection), it follows immediately from (4.3.3) that for given (t_0,x_0) the minimum of $V(t_0,x_0,u(\cdot))$ is

$$V_*(t_0,x_0) = x_0'P(t_0)x_0 \qquad (4.3.6)$$

if we may choose a controller u(t), $t_0 \leqslant t \leqslant T$ which, with its response x(t), $t_0 \leqslant t \leqslant T$ from $x(t_0) = x_0$ satisfies

$$u(t) + R^{-1}(t)B'(t)P(t)x(t) \equiv 0 \qquad t_0 \leqslant t \leqslant T. \quad (4.3.7)$$

As in Section 4.2.2, this requires $x(\cdot)$ to satisfy the homogeneous system

$$\dot{x}(t) = (A(t) - B(t)R^{-1}(t)B'(t)P(t))x(t), \; x(t_0) = x_0$$

from which $x(t)$ may be determined, and the required optimal controller is then given by (4.3.7). Once again, by (4.3.7), this optimal control law would be automatically generated by the feedback control law

$$u = -K(t)x \qquad (4.3.8)$$

where the time-varying feedback gain matrix is

$$K(t) = R^{-1}(t)B'(t)P(t). \qquad (4.3.9)$$

Thus, to summarize, provided the nonlinear Cauchy problem (4.3.4-5) has a solution $P(\cdot)$ on $[0,T]$, *the optimal control which minimizes* $V(t_0,x_0,u(\cdot))$ *for given* (t_0,x_0) *is automatically generated by the time-varying feedback law* (4.3.8-9), *and the minimum performance index is given by* (4.3.6).

4.3.3 The matrix Riccati differential equation

Equation (4.3.4) is the famous matrix Riccati differential equation, and we examine here the validity of the supposition in the previous subsec= tion that, with P prescribed at the final time T by (4.3.5), there exists a solution $P(\cdot)$ defined on the whole interval $[0,T]$. That this is not a trivial question is shown by Example 1.1.4, which exhibits the possibility of a solution of the Riccati-type equation (1.1.6) 'escaping' or 'blowing up' in finite time. The standard local existence theorem ensures that (4.3.4-5) has a solution $P(\cdot)$ defined on some interval $(t_1,T]$, and that this solution can fail to be further extendible (backwards in time) only if one or more entries of $P(t)$ are unbounded as $t \downarrow t_1$. By showing that this cannot happen we prove the following.

THEOREM 4.3.1 *The (unique) solution of the Riccati equation* (4.3.4) *with final time condition* (4.3.5) *extends backwards in time over the full interval* $[0,T]$.

PROOF Let $t_1 < T$ be an instant, as above, such that the solution $P(t)$ to (4.3.4-5) is defined on $(t_1,T]$. Then for $t_0 \in (t_1,T]$ our solution to the optimal control problem is valid, and in particular, (4.3.6) holds for any x_0. Since the performance index itself cannot take nega= tive values, it follows that

$$V_*(t_o,x_o) \geqslant 0 \quad \forall\ x_o \in \mathbb{R}^n, \quad t_1 < t_0 \leqslant T.$$

An upper bound for V_* is easily fixed by considering any particular con= troller, for example $u(t) \equiv 0$. For any x_0, $V(t_0,x_0,0)$ is defined and continuous in t_0 for $0 \leqslant t_0 \leqslant T$, and is consequently bounded as $t_0 \downarrow t_1$. Since

$$V_*(t_0,x_0) \leqslant V(t_0,x_0,0)$$

it follows that for each $x_0 \in \mathbb{R}^n$ the quantity

$$V_*(t_0,x_0) = x_0'P(t_0)x_0$$

is bounded both above and below as $t_0 \downarrow t_1$. From this result it fol= lows easily that each entry of $P(t_0)$ is bounded as $t_0 \downarrow t_1$, and hence the solution $P(\cdot)$ may be extended further back in time (unless $t_1 = 0$).■

REMARK It should be noted that this result is a consequence of the positive definiteness of $R(\cdot)$ and of the positive semidefiniteness of $Q(\cdot)$ and Q_f.

4.3.4 The general infinite-time problem

Finally we consider the system

$$\dot{x} = A(t)x + B(t)u \qquad 0 \leq t < \infty \qquad (4.3.10)$$

with performance index

$$V(t_0, x_0, u(\cdot)) = \int_0^\infty [u'(t)R(t)u(t) + x'(t)Q(t)x(t)]dt, \quad (4.3.11)$$

the assumptions on $A(\cdot)$, $B(\cdot)$, $R(\cdot)$, $Q(\cdot)$ being as in Section 4.3.1, with the additional condition that (4.3.10) be completely controllable (c.f. Section 1.3.1), so as to ensure the existence of at least one controller $u(\cdot)$ for which (4.3.11) is finite. The following theorem, which is fully proved in the text by Anderson and Moore (see References at the end of this lecture) shows that the solution to the optimal con= trol problem of minimizing (4.3.11) subject to (4.3.10) is just what one would expect upon letting $T \to \infty$ in the solution to the finite final time problem.

THEOREM 4.3.2 *Let* $P(t,T)$ $0 \leq t \leq T$ *denote the solution of the Riccati equation* (4.3.4) *with Cauchy condition*

$$P(T,T) = 0.$$

Then

$$\bar{P}(t) \triangleq \lim_{T \to \infty} P(t,T)$$

exists for all t, *is symmetric non-negative definite, and is also a solution of the Riccati equation* (4.3.4). *The optimal feedback law for* (4.3.10-11) *is*

$$u = -K(t)x = -R^{-1}(t)B'(t)\bar{P}(t)x$$

and the minimum value of $V(t_0, x_0, u(\cdot))$ *for given* $t_0 \geq 0$, x_0 *is*

$$V_*(t_0,x_0) = x_0'\bar{P}(t_0)x_0.$$

If A,B,R,Q are constant, it can be shown that $\bar{P}$ is also constant, and hence satisfies the *algebraic* Riccati equation (4.2.6).

4.3.5 The Hamiltonian system

If, in attempting to solve the optimal control problem of minimizing the performance index (4.3.2) subject to (4.3.1), one appeals to multiplier rules of the calculus of variations, one is led to the following linear homogeneous system of 2n differential equations in 2n variables, which appear as the *canonical* or *Hamiltonian* system of necessary conditions:

$$\begin{bmatrix}\dot{x}\\ \dot{\lambda}\end{bmatrix} = \begin{bmatrix}A(t) & -B(t)R^{-1}(t)B'(t)\\ -Q(t) & -A'(t)\end{bmatrix}\begin{bmatrix}x\\ \lambda\end{bmatrix} \qquad 0 \leq t \leq T \qquad (4.3.12)$$

$$x(t_0) = x_0, \qquad \lambda(T) = Q_f x(T). \qquad (4.3.13)$$

The connection between the Riccati equation and the Hamiltonian system is provided by the following theorem, the proof of which is simply a matter of direct verification, and is omitted.

THEOREM 4.3.3 Let $\begin{bmatrix}X(t)\\ \Lambda(t)\end{bmatrix}$ *be the* (2n×n) *matrix solution of the Cauchy problem*

$$\begin{bmatrix}\dot{X}\\ \dot{\Lambda}\end{bmatrix} = \left[\begin{array}{c|c}A(t) & -B(t)R^{-1}(t)B'(t)\\ \hline -Q(t) & -A'(t)\end{array}\right]\begin{bmatrix}X\\ \Lambda\end{bmatrix} \qquad 0 \leq t \leq T$$

$$\begin{bmatrix}X(T)\\ \Lambda(T)\end{bmatrix} = \begin{bmatrix}I_n\\ Q_f\end{bmatrix}.$$

Then X(t) *is invertible for all* $t \in [0,T]$, *and*

$$P(t) = \Lambda(t)X^{-1}(t)$$

is the solution of the Riccati equation (4.3.4) *with final time condi = tion* (4.3.5). *Conversely, having the solution* $P(\cdot)$ *to* (4.3.4-5), *the solution to the boundary value problem* (4.3.12-13) *is given by*

$$x(t) = \bar{x}(t), \quad \lambda(t) = P(t)\bar{x}(t), \quad t_0 \leqslant t \leqslant T$$

where $\bar{x}(\cdot)$ *is the solution of the linear homogeneous Cauchy problem*

$$\dot{\bar{x}} = (A(t) - B(t)R^{-1}(t)B'(t)P(t))\bar{x}, \quad t_0 \leqslant t \leqslant T$$

$$\bar{x}(t_0) = x_0.$$

REMARK There is a method involving the eigenvalues and eigenvectors of the Hamiltonian matrix for solving the algebraic Riccati equation (4.2.6) in the case of constant matrices A,B,R,Q. (c.f. the Anderson/ Moore text cited in the reference section below).

4.4 THE INVERSE LQ PROBLEM

The *inverse* LQ problem for the autonomous case treated in Section 4.2 is the following: given a completely controllable system

$$\dot{x} = Ax + Bu,$$

and an asymptotically stabilizing feedback law

$$u = -Kx, \tag{4.4.1}$$

is there a performance index

$$V(x_0,u(\cdot)) = \int_0^\infty (u'Ru+x'Qx)dt, \tag{4.4.2}$$

with R positive definite and Q positive semidefinite for which the control law (4.4.1) is optimal?

THEOREM 4.4.1 *For a given positive definite matrix* R, *there exists a positive semidefinite* Q *such that the given feedback law* (4.4.1) *is optimal for the performance index* (4.4.2) *iff for all real* ω *the matrix*

$$T^{\dagger}(i\omega)T(i\omega) - I$$

is positive semidefinite where $\dagger$ *denotes Hermitian conjugate, and where*

$$T(s) = I + R^{\frac{1}{2}}K(sI-A)^{-1}BR^{-\frac{1}{2}}$$ ■

4.5 THE DISCRETE-TIME LQ PROBLEM

For a discrete-time linear control system (c.f. Section 1.2.2)

$$x_{k+1} = A_k x_k + B_k u_k \qquad k=0,1,\ldots,N-1 \tag{4.5.1}$$

we might attempt to select a feedback law

$$u_i = -K_i x_i \qquad i=0,1,\ldots,N-1 \tag{4.5.2}$$

by minimizing a quadratic performance criterion of the form

$$V_i(x_i;u_1,\ldots,u_{N-1}) = \sum_{k=i}^{N-1}(u_k'R_k u_k + x_k'Q_k x_k) + x_N'Q_N x_N, \tag{4.5.3}$$

$$i=0,1,\ldots,N-1,$$

with each R_k positive definite and each Q_k positive semidefinite.

Although this is a finite-dimensional optimization problem, the same 'square-completion' device facilitates recursive solution.

For any symmetric matrices

$$P_0, P_1, \ldots, P_N,$$

we add the identity

$$x_N'P_Nx_N - x_i'P_ix_i = \sum_{k=i}^{N-1}(x_{k+1}'P_{k+1}x_{k+1} - x_k'P_kx_k)$$

$$= \sum_{k=i}^{N-1}[(A_kx_k+B_ku_k)'P_{k+1}(A_kx_k+B_ku_k) - x_k'P_kx_k]$$

to (4.5.3), and re-arrange to get

$$V_i(x_i,u_.) = x_i'P_ix_i + x_N'(Q_N-P_N)x_N$$
$$+ \sum_{k=i}^{N-1} [(u_k+\bar{R}_k^{-1}B_k'P_{k+1}A_kx_k)'\bar{R}_k(u_k+\bar{R}_k^{-1}B_k'P_{k+1}A_kx_k)$$
$$+ x_k'(A_k'P_{k+1}A_k-P_k+Q_k-A_k'P_{k+1}B_k\bar{R}_k^{-1}B_k'P_{k+1}A_k)x_k],$$

where

$$\bar{R}_k \triangleq R_k + B_k'P_{k+1}B_k. \qquad (4.5.4)$$

Hence if we determine the matrices $P_N, P_{N-1}, \ldots, P_o$ by backward recursion, starting with

$$P_N = Q_N$$

using (4.5.4) and

$$P_k = A_k'P_{k+1}A_k+Q_k - A_k'P_{k+1}B_k\bar{R}_k^{-1}B_k'P_{k+1}A_k, \qquad (4.5.5)$$

then V_i reduces to

$$V_i(x_i,u_.) = x_i'P_ix_i + \sum_{k=i}^{N-1} (u_k+\bar{R}_k^{-1}B_k'P_{k+1}A_kx_k)'\bar{R}_k(u_k+\bar{R}_k^{-1}B_k'P_{k+1}A_kx_k).$$

From this formula an easy induction shows that each matrix P_i is posi= tive semi-definite, and hence from (4.5.4) that $\bar{R}_k$ is also positive definite, and that the optimal feedback gain matrix is

$$K_i = \bar{R}_i^{-1}B_i'P_{i+1}A_i, \quad i=0,1,\ldots,N-1$$

with minimum performance index

$$V_{*i}(x_i) = x_i'P_ix_i, \quad i=0,1,\ldots,N-1.$$

For the infinite discrete-time case we only remark that the analogue of the algebraic Riccati equation is the matrix equation

$$P = A'PA + Q - A'PB[R+B'PB]^{-1}B'PA.$$

4.6 HISTORICAL NOTES AND REFERENCES

4.2.1-2 While quadratic functionals were studied intensively by Legendre, Jacobi and others as second variations in the calculus of variations, the importance of the linear-quadratic problem in control theory was first propagated by R.E. Kalman [1]. The 'square-completion' trick used here goes back to Legendre's transformation of the second variation - see, for example, Gelfand and Fomin [2]. The same answer to the optimal control problem can be derived in other ways, for example by dynamic programming. This approach was adopted by Anderson and Moore, in their highly recommended text [3].

4.2.6 Kleinman's iterative method was first presented in [4], and in [5] he gave a method for the automatic determination of an initial stabilizing feedback gain matrix K_1. Our convergence proof is adapted from K. Vitt [6]. Other methods of solving the algebraic Riccati equa= tion are described in [3].

4.3.2-4 This treatment follows that given in [3].

4.4 A discussion of the inverse problem is given in [3], and the theorem stated here is easily derived from the special case R = I as given by Molinari [7].

4.5 A full treatment of the discrete-time problem is given in the text by Kirk [8].

[1] R.E. Kalman, Contributions to the theory of optimal control. *Bol. Soc.Matem. Mex.*, pp. 102-119, 1960.

[2] I.M. Gelfand and S.V. Fomin, *Calculus of Variations*. Prentice-Hall, Inc., 1963.

[3] B.D.O. Anderson and J.B. Moore, *Linear Optimal Control*. Prentice-Hall, Inc., 1971.

[4] D.L. Kleinman, On an iterative technique for Riccati equation computations. *IEEE Trans. Aut. Control*, Vol AC-13, pp. 114-115, 1968.

[5] D.L. Kleinman, An easy way to stabilize a linear constant system. *IEEE Trans. Auto. Control* Vol AC-15, p. 692, 1970.

[6] K. Vitt, Iterative solution of the Riccati equation. *IEEE Trans. Aut. Control*, Vol AC-17, pp 258-259, 1972.

[7] B.Molinari, The stable regulator and its inverse. *IEEE Trans. Aut. Control*, Vol AC-18, pp 454-459, 1973.

[8] D.E. Kirk, *Optimal Control Theory - An Introduction*. Prentice-Hall, Inc., Englewood Cliffs, N.J., 1970.

Further suggestions

[9] P. Dorato and A. Levis, Optimal linear regulators: the discrete-time case. *IEEE Trans. Aut. Control*, Vol AC-16, pp 613-620, 1971.

[10] W.A. Coppel, Linear-quadratic optimal control. *Proc. Royal Soc. Edinburgh*, 73A, 18, 271-289, 1974/5.

[11] M.Athans, The role and use of the stochastic linear-quadratic-Guassian problem in control system design. *IEEE Trans. Aut. Control*, Vol AC-16, pp 529-552, 1971.

5. INTRODUCTION TO KALMAN FILTERING

5.1 INTRODUCTION

In the present lecture we address ourselves to the problem of extracting the information on the state of a linear system from the system's out= put measurements, in the presence of disturbances, both in the system itself and in the process of measuring the output.

Specifically, we shall consider the following discrete system

$$x(k+1) = Ax(k) + \Gamma w(k),\ x(0) = x_o; \qquad (5.1.1)$$

$$z(k) = Hx(k) + v(k). \qquad (5.1.2)$$

Here $x \in \mathbb{R}^n$ is the state vector, $z \in \mathbb{R}^m$ is the output vector, and the disturbances are the vectors $w \in \mathbb{R}^p$ and $v \in \mathbb{R}^q$. A, Γ and H are thus $n \times n$, $n \times p$ and $m \times n$ matrices, respectively. The discrete time-set is given by $k=0,1,2,\ldots$ and the disturbance sequences $w(k)$ and $v(k)$, $k=0,1,2,\ldots$ are independent Gaussian random variables with the following statistics, characteristic of so-called *white noise*:

$$E(w(k)) = E(v(k+1)) = 0,\ k=0,1,2,\ldots \qquad (5.1.3)$$

and covariance matrices

$$E(w(j)w'(k)) = \begin{cases} Q & \text{if } j=k \\ 0 & \text{if } j \neq k \end{cases} \qquad (5.1.4)$$

and

$$E(v(j+1)v'(k+1)) = \begin{cases} R & \text{if } j=k \\ 0 & \text{if } j \neq k. \end{cases} \qquad (5.1.5)$$

Note that the independence assumptions tell us that also

$$E(v(j)w'(k)) = 0 \text{ for all } j=1,2,\ldots \text{ and } k=0,1,2,\ldots \qquad (5.1.6)$$

Finally, the initial state x_0 is also a Gaussian random vector with the following statistics:

$$E(x_o) = 0, \tag{5.1.7}$$

$$E(x_0 x_0') = P_0, \tag{5.1.8}$$

and we shall assume in addition that the initial state is independent of the disturbances, i.e.

$$E(x_0\, w'(k)) = E(x_0 v'(k+1)) = 0,\ k=0,1,\ldots. \tag{5.1.9}$$

Thus the problem is posed of finding an estimate for x(k), given the output measurement sequence z(1), z(2)...,z(j); it is assumed through= out that the system statistics, given by (5.1.3)-(5.1.9), are known. We shall denote this estimate by $\hat{x}(k|j)$.

Specifically, our estimate $\hat{x}(k|j)$ will be determined by the requirement that the variance of the estimation error should be minimized, viz. $\hat{x}(k|j) = \arg\min_{\eta} (E((x(k)-\eta)'(x(k)-\eta)|z(1),\ldots z(j)))$. Obviously, $\hat{x}(k|j)$ is thus a function φ_k of the actual measurements,

i.e.

$$\hat{x}(k|j) = \varphi_k(z_1, z_2, \ldots, z_j). \tag{5.1.10}$$

If in (5.1.10) $k > j$, we then refer to the problem of *prediction*; if k=j we refer to the problem of *filtering*; while if $k < j$ one speaks about *smoothing*. In the present lecture we shall first treat the prediction problem, assuming that the filtering problem has been solved; we then turn to the filtering problem itself and obtain the full solution for the filtering and prediction problems. The smoothing problem is technically more involved, and we shall refrain from treating this problem here. In the final part of this lecture we comment on the problem of simultaneous control and estimation in the presence of disturbances, and present the now classical separation theorem.

5.2 GENERAL PROBABILISTIC BACKGROUND

If $X \in \mathbb{R}^k$ is a random vector, we denote by

$$F_X(x) = P(X \leqslant x) = P(X_1 \leqslant x_1, X_2 \leqslant x_2, \ldots X_k \leqslant x_k),\ x \in \mathbb{R}^k$$

the *probability distribution function* of X. Sometimes, when it is self-evident to which random vector we refer, we shall omit the sub= script X when referring to the probability distribution function.

If there exists a function $f(\xi_1, \xi_2, \ldots, \xi_k)$ such that

$$F(x) = \int_{-\infty}^{x_1} \cdots \int_{-\infty}^{x_k} f(\xi_1, \xi_2, \ldots, \xi_k) d\xi_1 \ldots d\xi_k \tag{5.2.1}$$

holds for all $x \in \mathbb{R}^k$, then the function $f(\cdot)$ is called the *probability density function* of the random (vector) variable X. Obviously

$$f(x) = \frac{\partial^k F}{\partial x_1 \ldots \partial x_k}$$

If we consider the probability of the specific event

$$A \triangleq \{X_1 \leqslant x_1, \ldots, X_\ell \leqslant x_\ell\},\ \text{where } \ell < k$$

viz.

$$F_X(x_1, x_2, \ldots, x_\ell, \infty, \ldots \infty) = P(A),$$

we then employ the notation

$$F_{X_1, \ldots, X_\ell}(x_1, \ldots, x_\ell) = P(A),$$

and refer to the *marginal probability density* function of $X_1, \ldots, X_\ell$. Obviously

$$F(x_1,\ldots,x_\ell) = \int_{-\infty}^{x_1} \text{-----} \int_{-\infty}^{x_\ell} \int_{-\infty}^{\infty} \text{----} \int_{-\infty}^{\infty} f(\xi_1,\ldots,\xi_k)d\xi_1\ldots d\xi_k \tag{5.2.2}$$

so that

$$F(x_1,\ldots x_\ell) = \int_{-\infty}^{x_1} \text{----}\int_{-\infty}^{x_\ell} f_{x_1\ldots x_\ell}(\xi_1\ldots\xi_\ell)d\xi_1\ldots d\xi_\ell \tag{5.2.3}$$

where

$$f_{x_1\ldots x_\ell}(x_1,\ldots x_\ell) = \int_{-\infty}^{\infty} \text{----} \int_{-\infty}^{\infty} f(x_1,\ldots x_\ell,\xi_{\ell+1}\ldots\xi_k)d\xi_{\ell+1}\ldots d\xi_k. \tag{5.2.4}$$

Here $f_{x_1\ldots x_\ell}$ is referred to as the marginal probability density func= tion of $X_1\ldots X_\ell$.

Next consider two events A and B. We denote by $P(A|B)$ the conditional probability of A occurring given that B has occurred:

$$P(A|B) \triangleq \frac{P(A\cap B)}{P(B)}$$

The function

$$F(x_1,\ldots,x_\ell|X_{\ell+1} = x_{\ell+1},\ldots X_k=x_k) \triangleq P(X_1 \leqslant x_1,\ldots X_\ell \leqslant x_\ell|\ X_{\ell+1}=x_{\ell+1},\ldots X_k=x_k)$$

is then called the conditional probability distribution function of $X_1,\ldots X_\ell$ given $X_{\ell+1},\ldots X_k$. Evidently the conditional probability density function of $X_1\ldots X_\ell$ given $X_{\ell+1},\ldots X_k$,

$$f(x_1,\ldots x_\ell|x_{\ell+1},\ldots x_k) \triangleq \frac{\partial^\ell F(x_1,\ldots,x_\ell|X_{\ell+1}=x_{\ell+1},\ldots,X_k=x_k)}{\partial x_1\ldots\partial x_\ell},$$

is then (in view of the definition of the conditional probability)

$$f(x_1,\ldots x_\ell|x_{\ell+1},\ldots x_k) = \frac{f(x_1,\ldots x_k)}{f(x_{\ell+1},\ldots x_k)} \quad . \tag{5.2.5}$$

Thus, note that if the joint probability density function $f(x_1,\ldots,x_k)$ is given, then the conditional probability density functions can be computed from (5.2.5) upon using (5.2.4); and the corresponding probability distribution function is computed by integrating the expression for the conditional probability density function, similar to (5.2.1)-(5.2.3).

We next introduce the definition of the expectation of a vector-valued function $h(\cdot)$ of the random vector X:

$$E(h(X)) = \int_{-\infty}^{\infty} \cdots \int_{-\infty}^{\infty} h(x)f(x)dx_1 \ldots dx_k.$$

In particular, the *mean* of the random vector X, denoted by $\bar{x}$, is

$$\bar{x} \triangleq E(X).$$

The expectation is a linear operator. The covariance matrix of the random vector X, say P, is the expectation of the matrix $(X-\bar{x})(X-\bar{x})'$, viz,

$$P \triangleq E((X-\bar{x})(X-\bar{x})') = \int_{-\infty}^{\infty} \cdots \int_{-\infty}^{\infty} (x-\bar{x})(x-\bar{x})'f(x)dx_1 \ldots dx_k.$$

It is thus readily verifiable that

$$P = \int_{-\infty}^{\infty} \cdots \int_{-\infty}^{\infty} xx'f(x)dx_1 \ldots dx_k - \bar{x}\bar{x}'. \qquad (5.2.6)$$

Similarly, the cross-covariance matrix of the random vectors $\tilde{X} \triangleq (X_1,\ldots X_\ell)$ and $\tilde{\tilde{X}} \triangleq (X_{\ell+1},\ldots X_k)$ is defined by

$$P_{\tilde{X}\tilde{\tilde{X}}} \triangleq E((\tilde{X}-\bar{\tilde{x}})(\tilde{\tilde{X}}-\bar{\tilde{\tilde{x}}})') = \int_{-\infty}^{\infty} \cdots \int_{-\infty}^{\infty} (\tilde{x}-\bar{\tilde{x}})(\tilde{\tilde{x}}-\bar{\tilde{\tilde{x}}})'f(x_1,\ldots,x_k)dx_1 \ldots dx_k.$$

Furthermore, as one would expect, the conditional expected value of the random vector $\tilde{X} = (X_1,\ldots,X_\ell)$ with respect to the random vector $\tilde{\tilde{X}} = (X_{\ell+1},\ldots,X_k)$ is

$$E_{\tilde{X}}(\tilde{X}|\tilde{\tilde{X}}) \triangleq \int_{-\infty}^{\infty} \ldots \int_{-\infty}^{\infty} \tilde{x}f(x_1,\ldots x_\ell|x_{\ell+1}\ldots x_k)dx_1,\ldots dx_\ell.$$

Similarly, the conditional covariance matrix is defined as

$$P_{\tilde{X}|\tilde{\tilde{X}}} \triangleq E_{\tilde{X}}((\tilde{X} - E(\tilde{X}|\tilde{\tilde{X}})(\tilde{X}-E(\tilde{X}|\tilde{\tilde{X}}))')$$

$$= \int_{-\infty}^{\infty} \cdots \int_{-\infty}^{\infty} (\tilde{x}-E(\tilde{X}|\tilde{\tilde{X}}))(\tilde{x}-E(\tilde{X}|\tilde{\tilde{X}}))'f(x_1,\ldots,x_\ell|x_{\ell+1},\ldots x_k)dx_1,\ldots dx_\ell.$$

Next we introduce the notion of independence. Two events A and B are said to be *independent* if

$$P(A \cap B) = P(A) \cdot P(B).$$

Thus, the elements $X_1, \ldots, X_\ell$ of the random vector X are independent if the probability distribution function can be written

$$F(x) = \prod_{i=1}^{k} F_i(x_i);$$

where

$$F_i(x_i) = P(X_i \leqslant x_i) \tag{5.2.7}$$

We shall also require the following definition.

The two random vectors $\tilde{X}$ and $\tilde{\tilde{X}}$ are *uncorrelated* if

$$E(\tilde{X}\tilde{\tilde{X}}') = E(\tilde{X})E(\tilde{\tilde{X}}').$$

Thus it should be noted that in view of (5.2.7), if the random vectors $\tilde{X}$ and $\tilde{\tilde{X}}$ are independent, then the vectors $\tilde{X}$ and $\tilde{\tilde{X}}$ are also uncorrelated; the converse is not true in general. Moreover, in view of (5.2.6) we conclude that if the vectors $\tilde{X}$ and $\tilde{\tilde{X}}$ are uncorrelated then the cross-co= variance matrix

$$P_{\tilde{X}\tilde{\tilde{X}}} = 0.$$

5.3 GAUSSIAN DISTRIBUTIONS

We shall mostly be interested in *Gaussian* probability distribution func= tions. Specifically, we say that the random vector X is Gaussian-distri= buted if its probability density function is of the form

$$f(x_1,\ldots x_k) = \frac{1}{\sqrt{(2\pi)^k|P|}} \exp(-\tfrac{1}{2}(x-\bar{x})'P^{-1}(x-\bar{x})), \tag{5.3.1}$$

where $\bar{x}$ is a fixed vector (which turns out to be the mean of X) and P is a positive definite symmetric matrix (with determinant $|P|>0$), which turns out to be the covariance matrix of the random vector X.

If the random vector X is split

$$X = \binom{\tilde{X}}{\tilde{\tilde{X}}},$$

we can then write the matrix P in (5.3.1) in the blocked form

$$P = \begin{bmatrix} P_1 & P_2 \\ P_3 & P_4 \end{bmatrix},$$

where

$$P_1 = P_{\tilde{X}\tilde{X}},\ P_2 = P_{\tilde{X}\tilde{\tilde{X}}},\ P_3 = P_{\tilde{\tilde{X}}\tilde{X}}\ (P_2 = P_3') \text{ and } P_4 = P_{\tilde{\tilde{X}}\tilde{\tilde{X}}}.$$

If we then write

$$P^{-1} = \begin{bmatrix} M_1 & M_2 \\ M_2' & M_4 \end{bmatrix},$$

it follows that

$$M_1 = (P_{\tilde{X}\tilde{X}} - P_{\tilde{X}\tilde{\tilde{X}}} P^{-1}_{\tilde{\tilde{X}}\tilde{\tilde{X}}} P_{\tilde{\tilde{X}}\tilde{X}})^{-1} = P^{-1}_{\tilde{X}\tilde{\tilde{X}}} P_{\tilde{\tilde{X}}\tilde{X}} M_4 P_{\tilde{\tilde{X}}\tilde{X}} P^{-1}_{\tilde{X}\tilde{\tilde{X}}}, \tag{5.3.2}$$

$$M_2 = -M_1 P_{\tilde{X}\tilde{\tilde{X}}} P_{\tilde{\tilde{X}}\tilde{\tilde{X}}}^{-1} = -P_{\tilde{X}\tilde{X}}^{-1} P_{\tilde{X}\tilde{\tilde{X}}} M_4, \tag{5.3.3}$$

$$M_4 = (P_{\tilde{X}\tilde{X}} - P_{\tilde{\tilde{X}}\tilde{X}} P_{\tilde{X}\tilde{X}}^{-1} P_{\tilde{X}\tilde{\tilde{X}}})^{-1} = P_{\tilde{\tilde{X}}\tilde{\tilde{X}}}^{-1} + P_{\tilde{\tilde{X}}\tilde{\tilde{X}}}^{-1} P_{\tilde{\tilde{X}}\tilde{X}} M_1 P_{\tilde{X}\tilde{\tilde{X}}} P_{\tilde{\tilde{X}}\tilde{\tilde{X}}}^{-1}. \tag{5.3.4}$$

We are now able to compute the conditional density distribution $f(x_1,\ldots x_\ell | x_{\ell+1} \ldots x_k)$.

To this end we first realize, in view of the formula (5.2.4) for the marginal density distribution function, and in view of (5.3.1), that the marginal density function is

$$f(x_{\ell+1},\ldots x_k) = \frac{1}{\sqrt{(2\pi)^{k-\ell}|P_{\tilde{\tilde{X}}\tilde{\tilde{X}}}|}}\exp(-\tfrac{1}{2}(\tilde{\tilde{X}}-\bar{\tilde{\tilde{x}}})'P_{\tilde{\tilde{X}}\tilde{\tilde{X}}}^{-1}(\tilde{\tilde{X}}-\bar{\tilde{\tilde{x}}})), \qquad (5.3.5)$$

i.e. it is also Gaussian.

Next, in view of the formula (5.2.5)(for the conditional probability density function), and of (5.3.2)-(5.3.4), and by direct manipulation we finally obtain

$$f(x_1,\ldots x_\ell|x_{\ell+1},\ldots x_k) = \frac{1}{\sqrt{(2\pi)^{\ell}|Q|}}\exp(-\tfrac{1}{2}(\tilde{x}-m)'Q^{-1}(\tilde{x}-m)), \quad (5.3.6)$$

where

$$m \triangleq \bar{\tilde{\tilde{x}}} + P_{\tilde{X}\tilde{\tilde{X}}}P_{\tilde{\tilde{X}}\tilde{\tilde{X}}}^{-1}(\tilde{\tilde{X}}-\bar{\tilde{\tilde{x}}}) \qquad (5.3.7)$$

and

$$Q \triangleq P_{\tilde{X}\tilde{X}}-P_{\tilde{X}\tilde{\tilde{X}}}P_{\tilde{\tilde{X}}\tilde{\tilde{X}}}^{-1}P_{\tilde{\tilde{X}}\tilde{X}}; \qquad (5.3.8)$$

thus $f(x_1,\ldots,x_\ell|x_{\ell+1},\ldots x_k)$ is a Gaussian density function with mean m (see (5.3.7)) and covariance Q(see (5.3.8)).

The following are two important properties of Gaussian random vectors.

LEMMA 5.3.1 *If* X *is a Gaussian random vector and* M *is an* $r \times k$ *matrix, then the random vector* MX *is Gaussian (with mean* $M\bar{x}$ *and covariance matrix* MPM'). *Moreover, the sum of the two Gaussian random vectors* $\tilde{X}$ *and* $\tilde{\tilde{X}}$ ($\tilde{X}$ *and* $\tilde{\tilde{X}}$ *are Gaussian in view of* 5.3.5) *is also a Gaussian random vector. Thus, loosely speaking, the Gaussian property is preserved under linear operations.* □

LEMMA 5.3.2 *Two random Gaussian vectors are uncorrelated if and only if they are independent.* □

5.4 THE CONDITIONAL EXPECTATION OF GAUSSIAN RANDOM VECTORS

LEMMA 5.4.1 $E(\tilde{X}|\tilde{\tilde{X}})$ *is a Gaussian random vector which is a linear combination of the elements of* $\tilde{\tilde{X}}$.

PROOF The lemma follows from the equation (5.3.7) for $E(\tilde{X}|\tilde{\tilde{X}})$ and from Lemma 5.3.1. □

LEMMA 5.4.2 *The random vector* $\tilde{X} - E(\tilde{X}|\tilde{\tilde{X}})$ *is independent of the random vector* $M\tilde{\tilde{X}}$ *where* M *is any* $r \times (k-\ell)$ *matrix.*

PROOF The two random vectors $\tilde{X} - E(\tilde{X}|\tilde{\tilde{X}})$ and $M\tilde{\tilde{X}}$ are Gaussian in view of Lemma 5.4.1 and 5.3.1, respectively. Hence in view of Lemma 5.3.2, it suffices to show that the cross-covariance matrix of these two random vectors is 0. Now

$$E(\tilde{X}-E(\tilde{X}|\tilde{\tilde{X}})) = E(\tilde{X}) - E_{\tilde{\tilde{X}}}E_{\tilde{X}}(\tilde{X}|\tilde{\tilde{X}})) = \bar{\tilde{x}} - \bar{\tilde{x}} = 0,$$

whereas

$$E(M\tilde{\tilde{X}}) = M\bar{\tilde{\tilde{x}}}.$$

Hence the cross-covariance matrix is

$$P_{\tilde{X}-E(\tilde{X}|\tilde{\tilde{X}}),M\tilde{\tilde{X}}} = E((\tilde{X}-E(\tilde{X}|\tilde{\tilde{X}}))(\tilde{\tilde{X}}-\bar{\tilde{\tilde{x}}})'M').$$

Thus, in view of (5.3.7):

$$P_{\tilde{X}-E(\tilde{X}|\tilde{\tilde{X}}),M\tilde{\tilde{X}}} = E((\tilde{X}-\bar{\tilde{x}}-P_{\tilde{X}\tilde{\tilde{X}}}P_{\tilde{\tilde{X}}\tilde{\tilde{X}}}^{-1}(\tilde{\tilde{X}}-\bar{\tilde{\tilde{x}}}))(\tilde{\tilde{X}}-\bar{\tilde{\tilde{x}}})'M')$$

$$= P_{\tilde{X}\tilde{\tilde{X}}}M' - P_{\tilde{X}\tilde{\tilde{X}}}P_{\tilde{\tilde{X}}\tilde{\tilde{X}}}^{-1}P_{\tilde{\tilde{X}}\tilde{\tilde{X}}}M' = P_{\tilde{X}\tilde{\tilde{X}}}M' - P_{\tilde{X}\tilde{\tilde{X}}}M' = 0,$$

so that we have independence. □

LEMMA 5.4.3 *If $\tilde{\tilde{X}}$ and $\tilde{\tilde{\tilde{X}}}$ are independent, then*

$$E(\tilde{X}|\tilde{\tilde{X}},\tilde{\tilde{\tilde{X}}}) = E(\tilde{X}|\tilde{\tilde{X}}) + E(\tilde{X}|\tilde{\tilde{\tilde{X}}}) - \bar{\tilde{x}} .$$

PROOF In view of equation (5.3.7) we have:

$$E(\tilde{X}|\tilde{\tilde{X}},\tilde{\tilde{\tilde{X}}}) = \bar{\tilde{x}} + P_{\tilde{X},\tilde{\tilde{X}}\tilde{\tilde{\tilde{X}}}} P_{\tilde{\tilde{X}}\tilde{\tilde{\tilde{X}}},\tilde{\tilde{X}}\tilde{\tilde{\tilde{X}}}}^{-1} \begin{bmatrix} \tilde{\tilde{X}}-\bar{\tilde{\tilde{x}}} \\ \tilde{\tilde{\tilde{X}}}-\bar{\tilde{\tilde{\tilde{x}}}} \end{bmatrix} . \tag{5.4.1}$$

Now

$$P_{\tilde{X},\tilde{\tilde{X}}\tilde{\tilde{\tilde{X}}}} = E\left((\tilde{X}-\bar{\tilde{x}}) \begin{pmatrix} \tilde{\tilde{X}}-\bar{\tilde{\tilde{x}}} \\ \tilde{\tilde{\tilde{X}}}-\bar{\tilde{\tilde{\tilde{x}}}} \end{pmatrix}'\right) = \left[P_{\tilde{X}\tilde{\tilde{X}}}, P_{\tilde{X}\tilde{\tilde{\tilde{X}}}}\right]$$

and

$$P_{\tilde{\tilde{X}}\tilde{\tilde{\tilde{X}}},\tilde{\tilde{X}}\tilde{\tilde{\tilde{X}}}} = E\left(\begin{pmatrix} \tilde{\tilde{X}}-\bar{\tilde{\tilde{x}}} \\ \tilde{\tilde{\tilde{X}}}-\bar{\tilde{\tilde{\tilde{x}}}} \end{pmatrix}((\tilde{\tilde{X}}-\bar{\tilde{\tilde{x}}})',(\tilde{\tilde{\tilde{X}}}-\bar{\tilde{\tilde{\tilde{x}}}})')\right)$$
$$= \begin{bmatrix} P_{\tilde{\tilde{X}}\tilde{\tilde{X}}} & 0 \\ 0 & P_{\tilde{\tilde{\tilde{X}}}\tilde{\tilde{\tilde{X}}}} \end{bmatrix} ,$$

because by hypothesis, the random vectors $\tilde{\tilde{X}}$ and $\tilde{\tilde{\tilde{X}}}$ are independent.

Hence, upon substituting into equation (5.4.1) we obtain

$$E(\tilde{X}|\tilde{\tilde{X}},\tilde{\tilde{\tilde{X}}}) = \bar{\tilde{x}} + \left[P_{\tilde{X}\tilde{\tilde{X}}}, P_{\tilde{X}\tilde{\tilde{\tilde{X}}}}\right] \begin{bmatrix} P_{\tilde{\tilde{X}}\tilde{\tilde{X}}}^{-1} & 0 \\ 0 & P_{\tilde{\tilde{\tilde{X}}}\tilde{\tilde{\tilde{X}}}}^{-1} \end{bmatrix} \begin{bmatrix} \tilde{\tilde{X}} - \bar{\tilde{\tilde{x}}} \\ \tilde{\tilde{\tilde{X}}} - \bar{\tilde{\tilde{\tilde{x}}}} \end{bmatrix}$$

$$= \bar{\tilde{x}} + P_{\tilde{X}\tilde{\tilde{X}}} P_{\tilde{\tilde{X}}\tilde{\tilde{X}}}^{-1}(\tilde{\tilde{X}}-\bar{\tilde{\tilde{x}}}) + P_{\tilde{X}\tilde{\tilde{\tilde{X}}}} P_{\tilde{\tilde{\tilde{X}}}\tilde{\tilde{\tilde{X}}}}^{-1}(\tilde{\tilde{\tilde{X}}}-\bar{\tilde{\tilde{\tilde{x}}}}).$$

Thus, again, in view of equation (5.3.7), we can write the above expres= sion in the form

$$E(\tilde{X}|\tilde{\tilde{X}},\tilde{\tilde{\tilde{X}}}) = E(\tilde{X}|\tilde{\tilde{X}}) + E(\tilde{X}|\tilde{\tilde{\tilde{X}}}) - \bar{\tilde{x}} .$$

□

Along similar lines it is also possible to prove the following result, which is a generalization of Lemma 5.4.3.

LEMMA 5.4.4 *Even when* $\tilde{\tilde{X}}$ *and* $\tilde{\tilde{\tilde{X}}}$ *are not independent, we have*

$$E(\tilde{X}|\tilde{\tilde{X}},\tilde{\tilde{\tilde{X}}}) = E(\tilde{X}|\tilde{\tilde{X}},\tilde{\tilde{\tilde{\tilde{X}}}}),$$

where

$$\tilde{\tilde{\tilde{\tilde{X}}}} = \tilde{\tilde{\tilde{X}}} - E(\tilde{\tilde{\tilde{X}}}|\tilde{\tilde{X}}) \tag{5.4.2}$$

and

$$E(\tilde{X}|\tilde{\tilde{X}},\tilde{\tilde{\tilde{\tilde{X}}}}) = E(\tilde{X}|\tilde{\tilde{X}}) + E(\tilde{X}|\tilde{\tilde{\tilde{\tilde{X}}}}) - \bar{\tilde{X}}. \tag{5.4.3}$$

5.5 GAUSSIAN-MARKOV STOCHASTIC PROCESSES

Consider the sequence $x(k)$, $k=0,1,2,\ldots$, which is the response of the system (5.1.1). Evidently in this case the sequence $x(k), k=0,1,2,\ldots$ is also a sequence of random vectors, and such sequences are commonly called (discrete) stochastic processes. For all $0 \leqslant j \leqslant k$ we have

$$x(k) = A^{k-j}\, x(j) + \sum_{i=j+1}^{k} A^{k-i}\, \Gamma w(i-1), \tag{5.5.1}$$

from which it is obvious that the stochastic process $x(k)$, $k=0,1,\ldots$ is Markovian (see Section 1.5 of Lecture 1). Moreover, upon setting $j=0$ in equation (5.5.1), we conclude in view of Lemma 5.3.1 that the stochastic process $x(k)$, $k=0,1,\ldots$ is Gaussian. Hence $x(k)$, $k=0,1,2,\ldots$ is a Gaussian-Markov stochastic process. In fact, we shall require the following properties.

The stochastic processes $x(k), k=0,1,\ldots$ and $z(i)$, $i=1,2,\ldots$ are Gaussian with identically zero means, and (5.5.2)

$$E(x(j)w'(k)) = 0 \text{ for all } k \geqslant j,\ j=0,1,\ldots; \tag{5.5.3}$$

$$E(z(j)w'(k)) = 0 \text{ for all } k \geqslant j,\ j=1,2,\ldots; \tag{5.5.4}$$

$$E(x(j)v'(k)) = 0 \text{ for all } j \text{ and } k, \text{ where } j=0,1,\ldots \text{ and } k=1,2,\ldots; \tag{5.5.5}$$

$E(z(j)v'(k)) = 0$ for all $k > j$, where $j, k = 1,2,\ldots.$ (5.5.6)

Indeed, it should be noted that in view of equation (5.1.3) and equations (5.1.1) and (5.1.2), $x(k)$ and $z(k)$, $k=0,1,\ldots$ have zero means.

Moreover, in view of (5.5.1), we know that

$$x(j) = A^j x_0 + \sum_{i=1}^{j} A^{j-i}\,\Gamma w(i-1)$$

$$= A^j x_0 + A^{j-1}\,\Gamma w(0) + A^{j-z}\,\Gamma w(1) + \ldots \Gamma\, w(j-1).$$

Upon post-multiplying this equation by $w'(k)$, and computing the expecta= tion, we obtain

$$E(x(j)w'(k)) = A^j E(x_0 w'(k)) + A^{j-1}\Gamma\, E(W(0)w'(k)) +$$

$$+ A^{j-2}\,\Gamma E(w(1)w'(k)) + \ldots \Gamma\, E(w(j-1)w'(k)).$$

In view of equation (5.1.9), the first term on the right-hand side of the above equation vanishes, and if $k \geqslant j$, then by (5.1.4) all the other terms on the right-hand side of this equation also vanish. Hence, if $k \geqslant j$, then $E(x(j)w'(k)) = 0$ and (5.5.2) holds. In order to verify (5.5.4) we set $j=k$ in equation (5.1.2) and post-multiply equation (5.1.2) by $w'(k)$. Taking expectations, we obtain

$$E(z(j),w'(k)) = H\, E(x(j)w'(k)) + E(v(j)w'(k)).$$

Now by (5.1.6) the second term on the right-hand side of the above equa= tion vanishes, and if $k \geqslant j$, then in view of (5.5.3) the first term on the right-hand side of this equation also vanishes, and therefore (5.5.4) holds. The equations (5.5.5) and (5.5.6) follow analogously.

5.6 OPTIMAL LEAST-MEAN-SQUARE ESTIMATION

Using the notation

$$z^*(j) \triangleq \begin{bmatrix} z(1) \\ z(2) \\ \vdots \\ z(j) \end{bmatrix} ;$$

and with the *estimation error* defined by

$$\tilde{x}(k|j) \triangleq x(k) - \hat{x}(k|j),$$

we have by (5.1.10)

$$\hat{x}(k|j) = \arg\min_{\eta} (E((x(k)-\eta)'(x(k)-\eta)|z^*(j)))$$

$$= \arg\min_{\eta} (E(\tilde{x}(k|j)'\tilde{x}\ (k|j)|z^*(j))),$$

and our mean-square-error loss function is $E(\tilde{x}(k|j)'\tilde{x}(k|j)|z^*(j))$.

The following simple result is crucial:

THEOREM 5.6.1 *The least-mean-square estimator is precisely the expected value conditioned on the observations - i.e.*

$$\hat{x}(k|j) = E(x(k)|z^*(j)).$$

PROOF Denote by $f(\xi)$ the probability density function of $x(k)$, conditioned on $z^*(j)$. Then the mean-square-loss function can be written as

$$E(\tilde{x}(k|j)'\tilde{x}(k|j)|z^*(j)) = \int_{-\infty}^{\infty} \text{----} \int_{-\infty}^{\infty} (\xi-\hat{x}(k|j))'(\xi-\hat{x}(k|j)f(\xi)d\xi_1\ldots d\xi_n$$

$$= \int_{-\infty}^{\infty} \text{----} \int_{-\infty}^{\infty} \xi'\xi f(\xi)d\xi_1\ldots d\xi_n + \hat{x}(k|j)'\hat{x}(k|j) - 2\hat{x}(k|j)'\bar{\xi}.$$

It is thus evident that the estimation $\hat{x}(k|j)$ which minimizes the above expression is

$$\hat{x}(k|j) = \bar{\xi}.$$

However, by definition, $\bar{\xi}$ is precisely $E(x(k)|z^*(j))$, i.e.

$$\hat{x}(k|j) = E(x(k)|z^*(j)). \qquad (5.6.1)$$

□

It is remarkable that the result (5.6.1) also holds for more general loss functions and for non-Gaussian probability density functions. For a detailed discussion, the reader is referred to the original paper by Kalman [1].

5.7 OPTIMAL PREDICTION

In this section we develop the optimal *prediction* formulae (i.e. those for $k > j$) assuming that the optimal filtering estimate $\hat{x}(j|j)$ and the associated covariance matrix of the filtering error

$$P(k|j) \triangleq E(\tilde{x}(k|j)\tilde{x}(k|j)'),$$

$k=j$, are known. This assumption is a prerequisite for obtaining the full solution for both the prediction and filtering problems in the next section. Note however that for $j=0$ we know:

$$\hat{x}(0|0) = E(x_0) = 0$$

and

$$P(0|0) = E(\tilde{x}(0|0)\tilde{x}(0|0)') = E(x_0 x_0') = P_0.$$

Indeed, the following will be shown to hold for all $k > j$.

THEOREM 5.7.1 *If the optimal filtered estimate* $\hat{x}(j|j)$ *and the covariance matrix* $P(j|j)$ *of the corresponding filtering error* $\tilde{x}(j|j) = x(j)-\hat{x}(j|j)$ *are known for some* $j = 0,1,\ldots$ *then for all* $k > j$:

(a) *the optimal predicted estimate* $\hat{x}(k|j)$, $k > j$, *is given by*

$$\hat{x}(k|j) = A^{k-j}\hat{x}(j|j); \qquad (5.7.1)$$

(b) *the prediction error stochastic process $\tilde{x}(k|j)$ is a zero-mean Gaussian-Markov sequence, the covariance matrix of which is governed by the relation*

$$P(k|j) = A^{k-j} P(j|j)A'^{k-j} + \sum_{i=j+1}^{k} A^{k-i} \Gamma Q\Gamma' A'^{k-i}. \quad (5.7.2)$$

PROOF (a) By Theorem 5.6.1 we know that

$$\hat{x}(k|j) = E(x(k)|z^*(j)) \equiv E(x(k)|z(1),z(2),\ldots z(j)).$$

Thus, in view of the equation (5.5.1) and by the linearity of the expectation operator, we have

$$\hat{x}(k|j) = A^{k-j}E(x(j)|z^*(j)) + \sum_{i=j+1}^{k} A^{k-i}\Gamma\, E(w(i-1)|z^*(j)). \quad (5.7.3)$$

Now from (5.5.4) we know that the two sets of random vectors $\{w(j), w(j+1),\ldots w(k-1)\}$ and $\{z(1),z(2),\ldots,z(j)\}$ are uncorrelated if $k-1 \geqslant j$, i.e. if $k \geqslant j+1$ (as is indeed the case). Since each of the vectors is Gaussian, this means that these two sets of random vectors are independent. Therefore

$$E(w(i-1)|z^*(j)) = E(w(i-1)) \text{ for all } i=j+1,j+2,\ldots k. \quad (5.7.4)$$

Finally, from (5.7.4) and in view of (5.1.3) we thus have

$$E(w(i-1)|z^*(j)) = 0 \text{ for all } i=j+1,j+2,\ldots k. \quad (5.7.5)$$

Upon substituting (5.7.5) into the right-hand side of (5.7.3) and noting that by Theorem 5.6.1, $E(x(j)|z^*(j)) = \hat{x}(j|j)$, we may conclude that (5.7.1) holds.

(b) The prediction error is

$$\tilde{x}(k|j) = x(k) - \hat{x}(k|j),$$

and in view of (5.5.1) we thus have

$$\tilde{x}(k|j) = A^{k-j}x(j) + \sum_{i=j+1}^{k} A^{k-i}\Gamma w(i-1)-\hat{x}(k|j).$$

Equation (5.7.1) then gives

$$\tilde{x}(k|j) = A^{k-j}\,\tilde{x}(j|j) + \sum_{i=j+1}^{k} A^{k-i}\,\Gamma w(i-1). \tag{5.7.6}$$

Next we shall show that

$$E(\tilde{x}(j|j)w'(i-1)) = 0 \text{ for } i=j+1,j+2,\ldots,k. \tag{5.7.7}$$

To this end we write

$$\tilde{x}(j|j) = x(j) - \hat{x}(j|j). \tag{5.7.8}$$

Now by (5.5.3) we have that

$$E(x(j),w'(i-1)) = 0 \text{ for } i=j+1,j+2,\ldots k. \tag{5.7.9}$$

Furthermore, we realize that in view of Theorem 5.6.1 (equation 5.6.1)) and in view of equations (5.3.7) and (5.5.2),

$$\hat{x}(k|j) = P_{x(k)z^*(j)}P^{-1}_{z^*(j)z^*(j)}z^*(j); \tag{5.7.10}$$

and, in particular, for k=j:

$$\hat{x}(j|j) = P_{x(j)z^*(j)}P^{-1}_{z^*(j)z^*(j)}z^*(j). \tag{5.7.11}$$

Equation (5.7.11) in turn implies that the optimal filtering estimate $\hat{x}(j|j)$ is a linear combination of the observations $z(1)$, $z(2),\ldots,z(j)$. Hence in view of equation (5.5.4) we conclude that also

$$E(\hat{x}(j|j)w'(i-1)) = 0 \text{ for all } i=j+1,j=2,\ldots,k. \tag{5.7.12}$$

Equations (5.7.8), (5.7.9) and (5.7.12) thus imply that equation (5.7.7) holds. This means that all the cross terms vanish that arise in the ex=

pected value of the right-hand side of equation (5.7.6) and its transpose. Therefore

$$P(k|j) = A^{k-j}P(j|j)A'^{k-j} + \sum_{i=j+1}^{k} A^{k-i}\,\Gamma Q\Gamma' A'^{k-i}.$$

Finally, it remains to be shown that $\tilde{x}(k|j)$ is a zero-mean Gaussian-Markov process. In this respect we have:

$$\tilde{x}(k|j) = x(k) - \hat{x}(k|j).$$

Upon substituting (5.7.1) and (5.5.1) into the above equation we have

$$\tilde{x}(k|j) = A^{k-j}\tilde{x}(j|j) + \sum_{i=j+1}^{k} A^{k-i}\,\Gamma w(i-1),$$

whence the zero-mean Gaussian property of $\tilde{x}(k|j)$ follows. In order to ascertain the Markov property of $\tilde{x}(k|j)$ we replace k by k+1 in the previous equation, obtaining,

$$\tilde{x}(k+1|j) = A^{k+1-j}\,\tilde{x}(j|j) + \sum_{i=j+1}^{k+1} A^{k+1-i}\Gamma w(i-1)$$

$$= A[A^{k-j}\tilde{x}(j|j) + \sum_{i=j+1}^{k} A^{k-i}\Gamma w(i-1)] + \Gamma w(k)$$

$$= A\tilde{x}(k|j) + \Gamma w(k),$$

which shows that the Markov property also holds. □

The following corollary is of importance in the subsequent development.

COROLLARY 5.7.2 *If the optimal filtered estimate* $\hat{x}(k|k)$ *and the covariance matrix* $P(k|k)$ *of the corresponding filtering error* $\tilde{x}(k|k)$ *are known for some* k=0,1,2... *then*

(a) *the single-stage optimal predicted estimate is given by the expression:*

$$\hat{x}(k+1|k) = A\,\hat{x}(k|k); \qquad (5.7.13)$$

(b) *the stochastic process* $\tilde{x}(k+1|k)$, k=0,1,... *(the single-stage prediction error) is a zero-mean Gaussian Markov process whose covariance matrix is given by the relation*

$$P(k+1|k) = AP(k|k)A' + \Gamma Q \Gamma'. \qquad (5.7.14) \quad \square$$

5.8 THE KALMAN FILTER

THEOREM 5.3.1 *Assume that the set of measurements* $\{z(1), z(2), \ldots z(k), z(k+1)\}$, *is given.*

(a) *The optimal filtered estimate* $\hat{x}(k+1|k+1)$ *is given by the recursive relation*

$$\hat{x}(0|0) = 0,$$

$$\hat{x}(k+1|k+1) = A\hat{x}(k|k) + K(k+1)[z(k+1) - HA\hat{x}(k|k)], \ k=0,1,.. \qquad (5.8.1)$$

(b) $$K(k+1) = P(k+1|k)H'[HP(k+1|k)H'+R]^{-1} \qquad (5.8.2)$$

$$P(k+1|k) = AP(k|k)A' + \Gamma Q \Gamma' \qquad (5.8.3)$$

and

$$P(k+1|k+1) = [I - K(k+1)H]P(k+1|k) \qquad (5.8.4)$$

for k=0,1,... *where* I *is the* $n \times n$ *identity matrix and* $P(0|0)=P_0$ *is the initial condition for equation* (5.8.3).

(c) *The filtering error stochastic process* $\tilde{x}(k+1|k+1)$, k=0,1,... *is a zero-mean Gaussian-Markov sequence whose covariance matrix is given by* (5.8.4).

PROOF (a) From Theorem (5.6.1) we know that the optimal filtered estimate is given by

$$\hat{x}(k+1|k+1) = E(x(k+1)|z(1),z(2),\ldots,z(k+1)). \qquad (5.8.5)$$

However, from equations (5.4.3) and (5.4.2) (and bearing in mind that $\bar{x}(k+1) = 0$) we obtain

$$\hat{x}(k+1|k+1) = E(x(k+1)|\, z(1), z(2),\ldots, z(k)) + E(x(k+1)|\, \tilde{z}(k+1|k)) \ ,(5.8.6)$$

$$= \hat{x}(k+1|k) + E(x(k+1)|\, \tilde{z}(k+1|k))$$

where

$$\tilde{z}(k+1|k) \quad = z(k+1) - E(z(k+1)|z(1),\ldots z(k)). \qquad (5.8.7)$$

REMARK Since $E(z(k+1)|z(1),\ldots z(k))$ is in fact (by Theorem 5.6.1) the optimal predicted estimate of $z(k+1)$ given the measurements $z(1)$, $\ldots z(k)$, we employ the notation

$$\hat{z}(k+1|k) = E(z(k+1)|z(1),z(2),\ldots z(k)), \quad (5.8.8)$$

and shall refer to

$$\tilde{z}(k+1|k) = z(k+1) - z(k+1|k)$$

as the *measurement residual*.

Substituting equation (5.1.2) into (5.8.8) we see that

$$\hat{z}(k+1|k) = HE(x(k+1)|\, z(1), z(2),\ldots z(k)) + HE(v(k+1)|\, z(1)\ldots, z(k)).$$

From equation (5.5.6), $v(k+1)$ is uncorrelated with each of the measurements $z(1),\ldots,z(k)$. Since all these vectors are Gaussian, it follows from Lemma 5.3.2 that $v(k+1)$ is independent of the set of measurements $z(1),\ldots z(k)$. Hence

$$E(v(k+1)|\, z(1),\ldots z(k)) = E(v(k+1)) = 0$$

(by (5.1.3)). Therefore

$$\hat{z}(k+1|k) = H\hat{x}(k+1|k). \tag{5.8.9}$$

Furthermore, from equation (5.7.13) we have

$$\hat{x}(k+1|k) = A\ \hat{x}(k|k) \tag{5.8.10}$$

so that

$$\hat{z}(k+1|k) = HA\ \hat{x}(k|k), \tag{5.8.11}$$

and the measurement residual is thus

$$\tilde{z}(k+1|k) = z(k+1) - HA\hat{x}(k|k). \tag{5.8.12}$$

Now, since x(k+1) and z(k+1|k) are zero-mean and Gaussian, upon employing equation (5.3.7) we obtain

$$E(x(k+1)|\,\tilde{z}(k+1|k)) = P_{x(k+1)\tilde{z}(k+1|k)}P^{-1}_{\tilde{z}(k+1|k)\tilde{z}(k+1|k)}\tilde{z}(k+1|k). \tag{5.8.13}$$

With the definition

$$K(k+1) = P_{x(k+1)\tilde{z}(k+1|k)}P^{-1}_{\tilde{z}(k+1|k)\tilde{z}(k+1|k)}, \tag{5.8.14}$$

and upon substitution of (5.8.13), (5.8.14), (5.8.10) and (5.8.12) into (5.8.6), it finally follows that the recursion (5.8.1) holds.

(b) In order to compute K(k+1) we must separately precompute the co= variance matrix $P_{\tilde{z}(k+1|k)\tilde{z}(k+1|k)}$ and the cross-covariance matrix $P_{x(k+1)\tilde{z}(k+1|k)}$. However, we see from (5.8.2) that K(k+1) is eventually expressed in terms of the covariance P(k+1|k); therefore we first have to express $\tilde{z}(k+1|k)$ in terms of $\tilde{x}(k+1|k)$. Indeed, in view of (5.8.9) the measurement residual can be expressed as

$$\tilde{z}(k+1|k) = z(k+1) - H\hat{x}(k+1|k). \tag{5.8.15}$$

Upon substituting (5.1.2) into (5.8.15) we thus obtain

$$\tilde{z}(k+1|k) = H\tilde{x}(k+1|k) + v(k+1).$$

Hence

$$P_{\tilde{z}(k+1|k)\tilde{z}(k+1|k)} = E((H\tilde{x}(k+1|k)+v(k+1))(H\tilde{x}(k+1|k)+v(k+1))')$$

$$= HP(k+1|k)H' + R + E(H\tilde{x}(k+1|k)v'(k+1))$$

$$+ E(v(k+1)(H\tilde{x}(k+1|k))').$$

Now the last two terms above vanish, and since

$$[E((H\tilde{x}(k+1|k)v(k+1)'))]' = E(v(k+1)(H\tilde{x}(k+1|k))')$$

it suffices to show that $E(H\tilde{x}(k+1|k)v'(k+1))$ vanishes.

We have

$$E(H\tilde{x}(k+1|k)v'(k+1)) = HE(\tilde{x}(k+1|k)v'(k+1))$$

and

$$E(\tilde{x}(k+1|k)v'(k+1)) = E(x(k+1)v'(k+1)) - E(\hat{x}(k+1|k)v'(k+1)). \quad (5.8.16)$$

From (5.5.5), the first term on the right-hand side of (5.8.16) vanishes. Furthermore, recalling (5.7.10), we see that the random vector $\hat{x}(k+1|k)$ is a linear combination of the random vectors $z(1), z(2), \ldots, z(j)$, and because the expectation operator is linear it follows from (5.5.6) that the second term on the right-hand side of (5.8.16) also vanishes for all $k=1,2,\ldots$; for $k=0$ recall that $\hat{x}(0|0) = 0$ so that also $\hat{x}(1|0) = A\hat{x}(0|0) = 0$. Hence, for all $k=0,1,2,\ldots$

$$P_{\tilde{z}(k+1|k)\tilde{z}(k+1|k)} = HP(k+1|k)H' + R. \quad (5.8.17)$$

Turning next to $P_{x(k+1)\tilde{z}(k+1|k)}$, we have

$$P_{x(k+1)\tilde{z}(k+1|k)} = E(x(k+1)\tilde{z}(k+1|k))$$

$$= E(x(k+1)\tilde{x}'(k+1|k))H' + E(x(k+1)v'(k+1)). \quad (5.8.18)$$

From equation (5.5.5) it is clear that the second term in (5.8.18) vanishes. Expanding the expectation in the first term in (5.8.18) we obtain

$$E(x(k+1)\tilde{x}'(k+1|k)) = E(\tilde{x}(k+1|k)\tilde{x}'(k+1|k)) + E(\hat{x}(k+1|k)\tilde{x}(k+1|k)). \quad (5.8.19)$$

Again, recalling that the random vector $\hat{x}(k+1|k)$ is a linear combina= tion of the random vectors $z(1)$, $z(2),\ldots,z(k)$, we conclude in view of Lemma 5.4.2 that

$$E(\hat{x}(k+1|k)\tilde{x}'(k+1|k)) = 0. \quad (5.8.20)$$

Substituting equations (5.8.20) and (5.8.19) into (5.8.18) we thus obtain

$$P_{x(k+1)\tilde{z}(k+1|k)} = P(k+1|k)H'. \quad (5.8.21)$$

Then substituting (5.8.21) and (5.8.17) into (5.8.14) we finally ob= tain the equation (5.8.2).

Equation (5.8.3) follows from Corollary (5.7.2), where the single-stage error-covariance matrix is given (see, for example, equation (5.7.14)).

In order to compute the covariance matrix of the filtering error, we shall first try to express the filtering error in terms of the one-step prediction error (see, for example, equation (5.8.4)). In this respect we have:

$$\tilde{x}(k+1|k+1) = x(k+1) - \hat{x}(k+1|k+1).$$

Substituting the expression for $\hat{x}(k+1|k+1)$ from (5.8.1) into the above equation, we obtain

$$\tilde{x}(k+1|k+1) = x(k+1) - A\hat{x}(k|k) - K(k+1)\tilde{z}(k+1|k). \quad (5.8.22)$$

Upon substituting into (5.8.22) equation (5.7.13) and the expression for $\tilde{z}(k+1|k)$ in the equation which follows equation (5.8.15), we obtain

$$\tilde{x}(k+1|k+1) = (I - K(k+1)H)\tilde{x}(k+1|k) - K(k+1)v(k+1). \quad (5.8.23)$$

The evaluation of $P(k+1|k+1) = E(\tilde{x}(k+1|k+1)\tilde{x}'(k+1|k+1))$

will involve the term $E(\tilde{x}(k+1|k)v'(k+1))$, which is the left-hand side of equation (5.8.16), and is therefore zero. Hence, upon computing $P(k+1|k+1)$ from equation (5.8.23) we obtain

$$\begin{aligned} P(k+1|k+1) = {} & (I-K(k+1)H)P(k+1|k)(I-K(k+1)H)' \\ & + K(k+1)RK'(k+1). \end{aligned} \quad (5.8.24)$$

Regrouping terms here and employing equation (5.8.2) for $K(k+1)$, we finally see that equation (5.8.4) holds.

(c) In the course of the proof of the Gaussian-Markov property of the prediction error in Theorem 5.7.1 we showed that

$$\tilde{x}(k|j) = A^{k-j}\tilde{x}(j|j) + \sum_{i=j+1}^{k} A^{k-i}\Gamma\, w(i-1). \quad (5.8.25)$$

Replacing k by $k+1$ and j by k in equation (5.8.25), we have

$$\tilde{x}(k+1|k) = A\tilde{x}(k|k) + \Gamma w(k). \quad (5.8.26)$$

Substituting (5.8.26) into our expression for $\tilde{x}(k+1|k+1)$ in terms of

$\tilde{x}(k+1|k)$, namely into (5.8.23), we obtain

$$\tilde{x}(k+1|k+1) = [I-K(k+1)H]A\tilde{x}(k|k) + [I-K(k+1)H]\Gamma w(k)-K(k+1)v(k+1),$$

whence the Gaussian-Markov property follows immediately. □

5.9 THE SEPARATION PRINCIPLE

We have so far considered the problem of getting an optimal estimate for the state of the uncontrolled system (5.1.1), (5.1.2). If we consider a controlled version of (5.1.1), viz.

$$x(k+1) = Ax(k) + Bu(k) +\Gamma\, w(k),\ x(0) = x_o, \quad (5.9.1)$$

where B is an $n\times p'$ matrix and the control sequence $u(k) \in \mathbb{R}^{p'}$ for $k=0,1,\ldots$; and if we assume that the control (input sequence) $u(k)$, $k=0,1,\ldots$ is given, it is then easy to show that the problem of optimal state estimation is basically unaltered. Specifically, the system state sequence $x(0), x(1),\ldots$ is a Gaussian-Markov process with the expectation of the state vector $\bar{x}(k)$ propagating in the obvious way ($\bar{x}(k+1) = A\bar{x}(k) + Bu(k)$) whereas the estimation error covariance is unaltered.

If in addition, one is given the performance index

$$J \triangleq \sum_{k=1}^{N} x'(k)Mx(k)+u'(k-1)Nu(k-1) \quad (5.9.2)$$

(here M and N are real symmetric and positive semidefinite $n\times n$ and $p'\times p'$ matrices, respectively), which is to be minimized, then in a deterministic setting, i.e. $\Gamma= 0$, (and assuming that the controller has access to the system state) the solution to the (deterministic) optimal control problem (5.1.1), (5.9.2) is given by

$$u(k) = F(k)x(k) \quad (5.9.3)$$

where the feedback matrix F(k) is determined recursively from

$$M(k+1) = S(k+1) + M \qquad (5.9.4)$$

$$F(k) = -[B'M(k+1)B+N]^{-1}B'M(k+1)A \qquad (5.9.5)$$

and

$$S(k) = A'M(k+1)A + A'M(k+1)BF(k), \qquad (5.9.6)$$

for $k=N-1,N-2,\ldots,0$, where $M(N) = M$.

In the stochastic case ($\Gamma \neq 0$) the system trajectory $x(k), k=0,1,\ldots,N$ is a Gaussian-Markov stochastic process and the performance index (5.9.2) is then replaced by the expectation of the performance index in (5.9.2), i.e.

$$y = E\left(\sum_{k=1}^{N} x'(k)Mx(k)+u'(k-1)Nu(k-1)\right). \qquad (5.9.3)$$

The optimal control problem (5.9.1), (5.9.2) and (5.9.3) is then con= sidered, and an optimal control sought of the form

$$u(k) = F_k(z^*(k),\bar{x}_o) \qquad (5.9.4)$$

where the unknown function F_k is a deterministic function of the measurement random vectors $\{z(1), z(2), \ldots, z(k)\}$ and of the vector $\bar{x}_o$,

It is thus remarkable that the following 'separation principle' holds.

THEOREM 5.9.1. *The optimal control system for the stochastic linear regulator* (5.9.1),(5.1.2) *and* (5.9.3) *consists of the optimal linear filter for the system* (5.1.1),(5.1.2) *cascaded with the optimal feedback gain matrix of the deterministic linear regulator* (5.9.1) (*with* $\Gamma=0$), (5.9.2). *Specifically,*

$$F_k(z^*(k),\bar{x}_0) = F(k)\hat{x}(k|k),$$

where $\hat{x}(k|k)$ *is constructed via* (5.8.1)-(5.8.4), *and* $F(k)$ *is given by* (5.9.4)-(5.9.6).

Thus, the two parts of the optimal control system ($\hat{x}(k|k)$ and $F(k)$) are determined separately; $\hat{x}(k|k)$ is determined by the system parameters $A,\Gamma,H,Q,R,P_0,\bar{x}_0$, whereas $F(k)$ is determined by the system parameters A,B,M,N. The following block diagram illustrates this point:

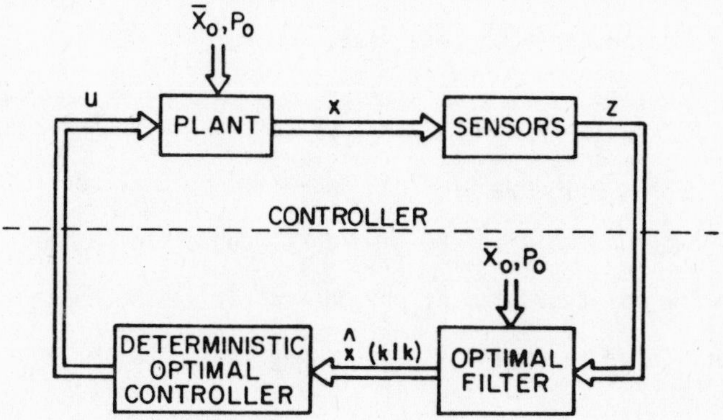

This remarkable result was first derived in [2] and [3].

5.10. REFERENCES

[1] Kalman, R.E., A new approach to linear filtering and prediction problems. *J. Basic Engng.*, Vol 82, pp 35-45, 1960.

[2] Joseph, P.D.; Tou, J.T., On linear control theory. *Trans AIEE*, pt. II, vol 80, pp 193, 1961.

[3] Gunckel, T.L.II; Franklin, G.F., A general solution for linear sampled data control. *J.Basic Engng.*, Vol 85, p 197, 1963.

Apart from these original papers on the Kalman filter and the separa= tion principle, the reader is referred to the excellent textbook:

Ho, Y.C.; Bryson, A.E., *Applied Optimal Control*. Halsted Press, 1975, Chapters 12-14.

5.11 CONTINUOUS-TIME KALMAN FILTERING

The problem statement in the continuous-time case requires the intro= duction of the difficult notion of a continuous-time stochastic process. The ensuing analysis of the continuous-time filtering problem and the proof of the separation principle for combined optimal filtering and control is then quite complicated and we shall therefore refrain from treating the continuous-time case here.

It is however remarkable that an analogy similar to that between the solutions to the continuous-time/discrete-time optimal control problems, also exists between the discrete-time/continuous-time optimal filtering problem and the joint problem of optimal filtering and control. In this connection we recommend the following textbooks:

Davis, M.H.A., *Linear Estimation and Stochastic Control*. Chapman and Hall, London, 1977.

Fleming, W.H., Rishel, W.R. *Deterministic and Stochastic Optimal Control*, Springer, New York, 1975.

and the hallmark paper:

Fugisaki, M., Kallianpur, G., Kunita, H., Stochastic Differential equations for the nonlinear filtering problem. *Osaka J. Math.*, Vol 9, 1973, pp 19-90.

6. THE MAXIMUM PRINCIPLE AND THE HAMILTON-JACOBI-BELLMAN EQUATION

6.1 BACKGROUND

Both the Maximum Principle of Pontryagin and the Hamilton-Jacobi-Bellman (H-J-B) Equation are used to characterize optimal solutions in nonlinear optimal control problems. In the linear-quadratic case one can dis= pense with these techniques and use only elementary algebra ('completing the square') to obtain results. However, extensions of the linear-quadratic formulation to include control constraints and nonlinearities produce more difficult problems which are solved by combining the ele= mentary techniques with aspects of Pontryagin and H-J-B theory. Accord= ingly, we first review this material before proceeding to extensions of the linear-quadratic problem.

Our attention will focus on the following optimal control problem: Mi= nimize with respect to all piecewise continuous r-vector functions of time $u(\cdot)$, and with respect to the final time T, the performance crite= rion

$$J(T,x_0,u(\cdot)) = \int_0^T L(t,x(t),u(t))dt + F(T,x(T)) \quad (6.1.1)$$

subject to the nonlinear dynamic system

$$\dot{x}(t) = f(t,x(t),u(t)), \quad x(0) = x_0, \quad (6.1.2)$$

the terminal constraint

$$\psi(T,x(T)) = 0 \quad (6.1.3)$$

and the control constraint

$$u(t) \in \Omega \subseteq R^r, \quad t \in [0,T]. \quad (6.1.4)$$

Note that the final time T is not specified, and is to be chosen along with u(·) to minimize J, and that ψ is an s-vector function of T and x.

As in Lecture 1 we assume that f is once continuously differentiable in x and u, and continuous in t (these assumptions may be weakened - see [1-4]), and that the same is true of L, F and ψ. Furthermore we assume that (6.1.2) has a unique solution for each piecewise continuous con= trol u(·).

6.2 THE PONTRYAGIN MAXIMUM PRINCIPLE

6.2.1 Statement of the maximum principle

THEOREM 6.2.1 [1-4] *Define* $H(t,x,u,\lambda) \triangleq \lambda_0 L(t,x,u) + \lambda' f(t,x,u)$. *Then, if* $u^*(\cdot)$, T^* *minimize* (6.1.1) *subject to* (6.1.2)-(6.1.4), *there exist a constant* $\lambda_0 \geq 0$, *a constant vector* $\nu \in R^s$, *and a continuous and piecewise continuously differentiable* n-*vector function of time* $\lambda(\cdot)$, *not all zero, such that for* $0 \leq t \leq T^*$

$$u^*(t) = \arg\min_{u \in \Omega} H(t,x^*(t),u,\lambda(t)) \qquad (6.2.1)$$

where

$$-\dot{\lambda}(t) = H_x'(t,x^*(t),u^*(t),\lambda(t)) \qquad (6.2.2)$$

and

$$\lambda(T^*) = \lambda_0 F_x'(T^*,x^*(T^*)) + \psi_x'(T^*,x^*(T^*))\nu. \qquad (6.2.3)$$

Moreover,

$$\lambda_0 F_T(T^*,x^*(T^*)) + \nu'\psi_T(T^*,x^*(T^*)) + H(T^*,x^*(T^*),u^*(T^*),\lambda(T^*)) = 0. \qquad (6.2.4)$$

If λ_0 can be set to unity, the problem is referred to as normal; it is a well-known fact, easy to prove, that if there are no terminal constraints (6.1.3) the problem is normal. If T is fixed, so that it may not be chosen to minimize J, the optimality condition (6.2.4) for T* falls away.

Pontryagin's Principle is a set of necessary conditions which must be satisfied by an optimal control. On the other hand, if a particular control function $\hat{u}(\cdot)$ satisfies Pontryagin's Principle it is only in special cases that one may conclude that $\hat{u}(\cdot)$ is optimal.

The proof of Pontryagin's Principle is involved; it introduced essentially new techniques in the Calculus of Variations as is evidenced by its use of the Brouwer fixed point theorem or even results of algebraic topology. When Ω is open, the Principle is easily reduced to a joint statement of the classical Euler/Lagrange, Weierstrass/Erdmann, transversality, and Weierstrass excess function necessary conditions.

EXAMPLE 6.2.2 We wish to control the linear dynamic system

$$\dot{x}(t) = Ax(t) + Bu(t), \quad x(0) = x_0 \qquad (6.2.5)$$

to minimize a linear function of the final x-vector, viz.

$$J = \alpha' x(T) \qquad (6.2.6)$$

subject to the constraint

$$-1 \leqslant u_i(t) \leqslant 1, \qquad i=1,\ldots,r. \qquad (6.2.7)$$

We assume that

$$\text{rank}\,[B_i, AB_i, \ldots, A^{n-1}B_i] = n, \quad i=1,\ldots,r, \qquad (6.2.8)$$

where B_i is the i-th column of B.

We then have the following theorem.

THEOREM 6.2.3 The controls

$$u_i^*(t) = -\text{sign}[B_i' e^{-A'(t-T)}\alpha], \quad i=1,\dots,r \qquad (6.2.9)$$

are well-defined and minimize $\alpha' x(T)$ *in the class of piecewise continuous controls that satisfy* (6.2.7).

PROOF First note that (6.2.8) ensures that (6.2.9) is well-defined, viz. $B_i' e^{-A'(t-T)}\alpha$ is not zero on a non-zero interval of time.

Applying Pontryagin's Principle, we see that

$$-\dot{\lambda}(t) = A'\lambda(t), \quad \lambda(T) = \alpha \qquad (6.2.10)$$

so that

$$H(t,x,u,\lambda) = \alpha' e^{-A(t-T)}(Ax+Bu) \qquad (6.2.11)$$

and H is minimized by (6.2.9). Although we have thus proved that (6.2.9) satisfies Pontryagin's Principle we have not proved that it actually minimizes $\alpha' x(T)$; this can be shown by a direct calculation which we leave to the reader, or by using the H-J-B equation as in Section 6.3.

6.2.2 The two-point boundary-value problem (TPBVP)

The above example is deceptively easy to treat because the conditions of Pontryagin's Principle can be solved explicitly. In general, how= ever, we have to solve the following pair of coupled equations with 'two-point' boundary conditions:

$$\dot{x}^*(t) = f(t,x^*(t),u^*(t)), \quad x(0) = x_0 \qquad (6.2.12)$$

$$-\dot{\lambda}(t) = H_x'(t,x^*(t),u^*(t)), \quad \lambda(T^*) = F_x'(T^*,x^*(T^*)) + \psi_x'(T^*,x^*(T))\nu. \qquad (6.2.13)$$

Here, not only are ν and T^* unknown, but the equations for x and λ can= not be integrated together because only some of the boundary conditions are known at t = 0 or t = T, viz. x(0) and not $\lambda(0)$ at t = 0, and not x(T) at t = T. It turns out that in the linear-quadratic problem with T specified, the TPBVP can be solved by setting

$$\lambda(t) = P(t)x(t) \qquad (6.2.14)$$

where $P(\cdot)$ satisfies the matrix Riccati differential equation.

6.3 H-J-B EQUATION: SUFFICIENCY

As pointed out in Section 6.2, Pontryagin's Principle is a set of neces= sary conditions for optimality. On the other hand, the H-J-B theory offers a sufficient condition; viz. if a solution can be found to the H-J-B equation, then the optimal control can be computed directly and easily from that solution. The catch is that it is not easy in general to solve the H-J-B equation and one has to resort to numerical techniques.

We first consider the case of no terminal constraints and fixed T. The H-J-B equation is then

$$-V_t(t,x) = \min_{u \in \Omega} [\, L(t,x,u)+V_x(t,x)f(t,x,u)] \qquad (6.3.1)$$

$$V(T,x) = F(T,x), \qquad (6.3.2)$$

and we have the following theorem.

THEOREM 6.3.1 *Suppose that there exists a once continuously differen= tiable scalar function* $V(\cdot,\cdot)$ *of* t *and* x *that satisfies* (6.3.1-2). *Suppose further that the control* $u^*(t,x(t))$ *that minimizes*

$$L(t,x(t),u(t)) + V_x(t,x(t))f(t,x(t),u(t)) \qquad (6.3.3)$$

subject to the constraint $u(t) \in \Omega$ *is such that* (6.1.2) *has a solution defined on* $[0,T]$ *and that the resulting control* $u^*(t,x^*(t))$ *is piece= wise continuous in* t *on* $[0,T]$. *Then,* $u^*(\cdot,x^*(\cdot))$ *minimizes* (6.1.1) *and* $V(0,x_0)$ *is equal to the minimum value of* $J(T,x_0,u(\cdot))$.

PROOF Under the stated assumptions we have

$$V(0,x(0)) - V(T,x(T)) + \int_0^T \frac{d}{dt} V(t,x(t))dt = 0. \quad (6.3.4)$$

Adding this identically zero quantity to (6.1.1), noting that

$$\frac{d}{dt} V(t,x(t)) = V_t(t,x(t)) + V_x(t,x(t))f(t,x(t),u(t)) \quad (6.3.5)$$

and supposing that $V(t,x(t))$ satisfies (6.3.1-2), we have

$$J(T,x_0,u(\cdot)) = V(0,x_0) + \int_0^T [H(t,x(t),u(t),V_x(t,x(t)))$$

$$- H(t,x(t),u^*(t,x(t)),V_x(t,x(t)))]dt \qquad (6.3.6)$$

where

$$H(t,x,u,V_x) \triangleq L(t,x,u) + V_x f(t,x,u). \qquad (6.3.7)$$

Now the integrand in (6.3.6) is non-negative and takes on its minimum value of zero when $u(t) = u^*(t,x(t))$, so that the theorem is proved.

EXAMPLE 6.3.2 We apply the H-J-B equation to Example 6.2.2. Here the H-J-B equation is

$$-V_t(t,x) = \min_{u \in \Omega} [V_x(t,x)Ax+V_x(t,x)Bu] \qquad (6.3.8)$$

which yields

$$u_i^*(t,x(t)) = -\text{sign}[B_i' V_x'(t,x(t))], \quad i=1,\ldots,r \tag{6.3.9}$$

and

$$-V_t(t,x) = V_x(t,x)Ax - \sum_{i=1}^{r} |V_x(t,x)B_i|, \tag{6.3.10}$$

$$V(T,x) = \alpha' x. \tag{6.3.11}$$

Now the choice

$$V(t,x) \triangleq \alpha' e^{-A(t-T)} x - \int_t^T \sum_{i=1}^{r} |\alpha' e^{-A(\tau-T)} B_i| d\tau \tag{6.3.12}$$

can by direct substitution be shown to satisfy (6.3.10), and from (6.3.12) we see that

$$V_x(t,x) = \alpha' e^{-A(t-T)}. \tag{6.3.13}$$

Substituting (6.3.13) into (6.3.9) we have (6.2.9), which proves its optimality.

When terminal constraints are present (we assume T fixed) the situation is more complicated. The H-J-B equation becomes

$$-V_t(t,x) = \min_{u \in \Omega} [L(t,x,u)+V_x(t,x)f(t,x,u)] \tag{6.3.14}$$

$$V(T,x) = F(T,x), \text{ for all } x \text{ such that } \psi(T,x) = 0 \tag{6.3.15}$$

$V(T,x)$ *unspecified* for all x such that $\psi(T,x) \neq 0$.

We then have the following generalization of Theorem 6.3.1.

THEOREM 6.3.3 Suppose there exists a once continuously differentiable function $V(\cdot,\cdot)$ *of* t *and* x *that satisfies* (6.3.14-15). *Suppose further that* $u^*(t,x(t))$ *minimizes*

$$L(t,x(t),u(t)) + V_x(t,x(t))f(t,x(t),u(t)) \qquad (6.3.16)$$

subject to $u(t) \in \Omega$, *causes* (6.1.1) *to have a solution defined on* $[0,T]$, $u^*(\cdot,x^*(\cdot))$ *to be piecewise continuous, and*

$$\psi(T,x^*(T)) = 0. \qquad (6.3.17)$$

Then, $u^*(\cdot,x^*(\cdot))$ *minimizes* (6.1.1)*and* $V(0,x_0)$ *is equal to the minimum value of* $J(T,x_0,u(\cdot))$.

PROOF The proof is very similar to that of Theorem 6.3.1 and is left as an exercise for the reader.

EXAMPLE 6.3.4 Consider the problem of minimizing

$$J = \int_0^T u'(t)u(t)dt \qquad (6.3.18)$$

subject to

$$\dot{x}(t) = A(t)x(t) + B(t)u(t), \qquad (6.3.19)$$

and

$$x(0) = x_0, \quad x(T) = 0, \qquad (6.3.20)$$

where it is assumed that (6.3.19) is completely controllable on $[0,T]$.

If we set

$$V(t,x) = 2x'\Phi'(T,t)W^{-1}(0,T)\Phi(T,0)x_0 - \int_t^T x_0'\Phi'(T,0)W^{-1}(0,T)\Phi(T,t)B(t)B'(t)\Phi'(T,t)W^{-1}(0,T)\Phi(T,0)x_0 dt \qquad (6.3.21)$$

we see that at $t = T$, $V(T,x) = 0$ when $x = 0$. As $F(T,x) \equiv 0$, this is as required by (6.3.15).

The function (6.3.21) is easily seen to satisfy (6.3.14), and $u^*(t,x(t))$

turns out as

$$u^*(t,x(t)) = -B'(t)\Phi'(T,t)W^{-1}(0,T)\Phi(T,0)x_0. \qquad (6.3.22)$$

A direct calculation, using the expression for the solution of (6.3.19) given in Lecture 1, shows that (6.3.19) driven by (6.3.22) yields $x(T) = 0$, as desired.

By Theorem 6.3.3 we therefore have that (6.3.22) is the optimal control, and from (6.3.21) the minimum value of (6.3.18) is

$$V(0,x_0) = x_0'\Phi'(T,0)W^{-1}(0,T)\Phi(T,0)x_0. \qquad (6.3.23)$$

Naturally, in more complicated problems it is more difficult to find an appropriate solution of (6.3.14-15); however, this is the price one pays for attempting to derive a globally optimal control function for nonlinear problems with terminal constraints.

We indicated in Section 6.2.1 that Pontryagin's Principle can be solved in the linear-quadratic case by assuming a linear relationship (6.2.14) between $\lambda(t)$ and $x(t)$. Similarly, the H-J-B equation can be solved in the linear-quadratic case by assuming that $V(t,x)$ is a quadratic expres= sion in x. As this is well known we mention it here only for com= pleteness and do not present any details.

6.4 COMPUTATIONAL ASPECTS

In general it is not possible to directly solve either the TPBVP ari= sing from Pontryagin's Principle or the H-J-B equation; consequently one has to resort to numerical methods.

In the case of the TPBVP one uses successive approximation methods to improve a nominal guessed solution; if boundary conditions are guessed, the methods are known as shooting methods; if control func= tions are successively improved, the methods are known as gradient-type methods in function space; and if both state and control functions are successively improved, the methods are known as quasi-linearization methods. Numerical approaches to the H-J-B equation include global solution by discretization of the partial differential equation in both state and time coordinates and successive improvement of a guessed con= trol function by exploiting expansions of the H-J-B equation in the neighbourhood of the guessed control function. The latter approach is known as Differential Dynamic Programming (DDP).

The above-mentioned approaches are summarized in [5], where DDP is also developed in detail. Reference [5] is now dated in some respects and accordingly we refer the reader to [6-8] for a variety of improvements and developments in numerical methods for computing optimal controls.

6.5 REFERENCES

[1] E.B. Lee and L. Markus, *Foundations of optimal control theory*. Wiley, New York, 1967.

[2] P.P. Varaiya, *Notes on optimization*. Van Nostrand, New York, 1972.

[3] G.F. Bryant and D.Q. Mayne, The maximum principle. *International Journal of Control*, Vol 20, 1974, pp 1021-1054.

[4] H.Halkin, Mathematical foundations of system optimization. Chap= ter 6 of *Topics in Optimization*, G. Leitmann (ed.), Academic Press, New York, 1967.

[5] D.H. Jacobson and D.Q. Mayne, *Differential dynamic programming*. Elsevier, New York, 1970.

[6] IEEE Transactions on Automatic Control: 1969-1979.

[7] Journal of Optimization Theory and Applications: 1969-1979.

[8] International Journal of Control: 1969-1979.

ADDITIONAL BIBLIOGRAPHY

L.S. Pontryagin, V.G. Boltyanskii, R.V. Gamkrelidze and E.F. Mischenko, *The mathematical theory of optimal processes*. English translation edited by L.W. Neustadt, Wiley Interscience, New York, 1962.

M.R. Hestenes, *Calculus of Variations and Optimal Control*. Wiley, New York, 1966.

V.G. Boltyanskii, The method of tents in the theory of extremum pro= blems. English Trans. in *Russian Math. Surveys*, 30(3), 1975, pp. 1-54.

7. THE NON-CONVEX CASE

7.1 INTRODUCTION

In this lecture we consider the system

$$\dot{x} = Ax + Bu, \quad 0 \leqslant t \leqslant T, \quad x(0) = x_0 \tag{7.1.1}$$

and the cost functional

$$J(T,x_0,u(\cdot)) = \int_0^T (x'Qx+u'Ru)dt + x'(T)Q_f x(T). \tag{7.1.2}$$

Here the vectors x, $x_0 \in \mathbb{R}^n$, $u \in \mathbb{R}^m$ and A,B,R,Q,Q_f are real matrices with the proper dimensions ; in addition, the matrices R,Q and Q_f are symmetric. The problem is posed of minimizing the performance functional (7.1.2) subject to the constraint (7.1.1).

In standard linear-quadratic theory it is assumed that the matrix R is positive-definite (one can then, without loss of generality, replace R by the unit matrix), that the matrix Q is non-negative definite, i.e. $Q = C'C$ for some $r \times n$ matrix C, and that $Q_f = 0$. One then has a convex optimization problem, and for the case $T \to \infty$(the regulator problem) the following classical results hold - c.f. Theorem 1.

THEOREM 7.1.1 (Kalman 1961 [1]) *If the pair* (A,B) *is controllable then there exists a solution* $u^*(\cdot)$ *to the infinite-time optimal control problem,*

$$u^* = -R^{-1}B'Px \tag{7.1.3}$$

and

$$J(\infty,x_0,u^*(\cdot)) = x_0'Px_0 \; ; \tag{7.1.4}$$

here P *is a real symmetric non-negative definite solution of the algebraic matrix Riccati equation* (ARE)

$$A'P+PA-PBR^{-1}B'P+Q = 0. \qquad (7.1.5)$$

Moreover, if the pair (A,C) *is observable, the matrix* P *is the unique positive definite solution of the* ARE (7.1.5) *and the optimal closed-loop system is asymptotically stable.* □

We shall next require the following definition.

DEFINITION *We say that the system* (7.1.1)(or: the pair (A,B)) *is stabilizable if there exists a feedback* F *such that the closed-loop system* (7.1.1), *i.e. the homogeneous system*

$$\dot{x} = (A+BF)x \quad , \quad x(0) = x_0,$$

is asymptotically stable. Similarly, the system

$$\dot{x} = Ax, \quad x(0) = x_0$$
$$y = Cx$$

is detectable if the pair (A',C') *is stabilizable.*

We are now able to state the following theorem (c.f. Theorem 4).

THEOREM 7.1.2 (Wonham 1968 [2]) *In the hypotheses of Theorem* 7.1.1 *one may replace controllability by stabilizability, and observability by detectability.* □

In the present lecture we shall relax the assumptions of standard li= near-quadratic theory that Q is non-negative definite and $Q_f = 0$. Specifically, in Section 7.2 we discuss ordering, stability properties and counting of the solutions to the ARE; whenever possible we also relax the assumption of controllability of the pair (A,B). In Section 7.3 we consider the relationship of this approach to that of Potter [3].

Further properties of the solution to the ARE are given in Section 7.4. We also comment on the connection with the regulator problem. Know=ledge of the solutions of the ARE proves useful in the discussion of the Riccati differential equation (the RE)

$$\frac{dP(t)}{dt} = A'P(t) + P(t)A + Q - P(t)BR^{-1}B'P(t), \; P(0) = Q_f, \quad (7.1.6)$$

in Section 7.5.

Several bounds on the solution to the RE are given, the 'regions of attraction' of solutions to the ARE are considered and we also comment on the stability of the initial matrix Q_f. The existence of a solution to the RE (7.1.6) is intimately related to the question of conjugate/focal times in the associated optimal control problem (7.1.1)(7.1.2) (or in the optimal control problem whose linearization gives rise to (7.1.1) and (7.1.2)).

We assume that the matrix R is positive definite, except in Section 7.6, where we let R be indefinite, as is the case in the algebraic Riccati equation which arises in infinite-time linear-quadratic zero-sum dif=ferential games.

We shall here refrain from including lengthy proofs, and refer the reader to the relevant literature; instead we present short derivations of results, and examples, and discuss the implications of the import=ant theorems.

We denote by $\chi^-(A)$, $\chi^+(A)$ and $\chi^c(A)$ the subspaces which are spanned by all the eigenvectors and generalized eigenvectors of A which correspond to the eigenvalues of A with negative, non-negative and zero real parts respectively.

It is well known [4] that these subspaces are invariant under A, and in addition

$$\chi^+(A) \oplus \chi^-(A) = \mathbb{R}^n. \qquad (7.1.7)$$

Furthermore, the pair (A,B) is stabilizable iff

$$\chi^+(A) \subset \text{Range}\,([B \,\vdots\, AB \,\vdots\, \text{----} \,\vdots\, A^{n-1}B]). \qquad (7.1.8)$$

Similarly, the pair (A,C) is detectable iff

$$(R([C' \,\vdots\, A'C' \,\vdots\, \text{----} \,\vdots\, A'^{n-1}C']))^{\perp} \subset \chi^-(A).$$

We shall denote by $(\nu)_A$ the largest subspace of the vector space $\nu \subset \mathbb{R}^n$ invariant under A, and

$\text{Ker}(A) \triangleq$ null space or kernel of A, i.e. $\{x|Ax = 0\}$.

Finally, the set of all the real symmetric solutions to the ARE will be denoted by S.

7.2 CLASSIFICATION OF SOLUTIONS

The convex cone of all n×n symmetric positive semidefinite matrices defines a natural partial ordering:

$$P_1 \leqslant P_2 \text{ iff } P_2 - P_1 \text{ is positive semidefinite.}$$

THEOREM 7.2.1 [5] *Assume that* S *is non-empty. Then the pair* (A,B) *is stabilizable iff there exists a unique* $P^+ \in S$ *such that* $\text{Re}(\lambda(A-BR^{-1}B'P^+)) \leqslant 0$ *and* $P \leqslant P^+$ *for* P *any element of* S. □

Theorem 7.2.1 shows that if the pair (A,B) is stabilizable, all real symmetric solutions of the ARE may be ordered with respect to a unique solution P^+, and we refer to P^+ as the *maximal solution* to the ARE. This result is related to Theorem 5 in [6] which states that if the system

(7.1.1) is controllable and the following frequency domain inequality is satisfied

$$H(-j\omega,\ j\omega) \geqslant 0 \qquad (7.2.1)$$

for all real ω, where

$$H(\bar{s},s) \triangleq R + B'(I\bar{s}-A')^{-1}Q(Is-A)^{-1}B,$$

then there is a unique $P = P^{+} \in S$ with the properties listed in Theorem 7.2.1. Theorem 5 of [6] also asserts that if the pair (A,B) is controllable, there is also a *minimal solution* P^{-}. However, stabilizability is not sufficient to ensure the existence of a P^{-}, as shown in the following example.

EXAMPLE 7.2.2 Let $A = \begin{bmatrix} 2 & 1 \\ 0 & -1 \end{bmatrix}$, $B = \begin{bmatrix} 1 \\ 0 \end{bmatrix}$, $R = \begin{bmatrix} 1 & 0 \\ 0 & 1 \end{bmatrix}$, $Q = \begin{bmatrix} -3 & -1 \\ -1 & 0 \end{bmatrix}$.

Then $P^{+} = \begin{bmatrix} 3 & 1 \\ 1 & \frac{1}{2} \end{bmatrix}$

and the other solutions to the ARE are given by

$$P_{\alpha} = \begin{bmatrix} 1 & \alpha \\ \alpha & \alpha-\alpha^2/2 \end{bmatrix}, \quad \alpha \in R^1.$$

For all α, $P_{\alpha} \leqslant P^{+}$, $\mathrm{Re}(\lambda(A-BR^{-1}B'P^{+})) = \mathrm{Re}(\lambda(\begin{smallmatrix} -1 & 0 \\ 0 & -1 \end{smallmatrix})) < 0$ as expected; however there is no $\alpha = \alpha^{*}$ for which $P_{\alpha^{*}} \leqslant P_{\alpha}$ for all $\alpha \in R^2$.

The discussion is concluded by the following theorem, which shows that stabilizability of (-A,B) is a necessary and sufficient condition for the existence of P^{-}.

THEOREM 7.2.3 *Assume that* S *is non-empty. Then the pair* (-A,B) *is stabilizable iff there exists a unique* $P = P^{-} \in S$ *such that* $\mathrm{Re}(\lambda(A-BR^{-1}B'P^{-})) \geqslant 0$ *and* $P \geqslant P^{-}$ *for all* $P \in S$.

PROOF Apply Theorem 7.2.1 to $(-A)'K + K(-A) + Q - KBR^{-1}B'K = 0$. From the theorem, this equation has a unique $K = K^{+}$ such that $\mathrm{Re}(\lambda(-A-BR^{-1}B'K^{+})) \leqslant$ and $K \leqslant K^{+}$ for all other solutions K. Now let $P = -K$, so that P satis=

fies the usual Riccati equation which, by the preceding, must also have a unique solution $P = P^- = -K^+$ such that $-P \leqslant -P^-$ or $P^- \leqslant P$ for all $P \in S$ and $Re(\lambda(-A-BR^{-1}B'(-P^-))) \leqslant 0$ or $Re(\lambda(A-BR^{-1}B'P^-)) \geqslant 0$

For the converse, define $K = -P$ and $K^+ = -P^-$, so that $(-A)'K + K(-A) + Q - KBR^{-1}B'K = 0$ has a unique solution $K = K^+$ such that $Re(\lambda(-A-BR^{-1}B'K^+)) \leqslant 0$ and $K \leqslant K^+$ for any other solution. By Theorem 7.2.1, $(-A,B)$ is stabilizable. □

Theorems 7.2.1 and 7.2.3 yield the following corollary, which also gives the result of Willems [6] that controllability of (A,B) is suf= ficient for the existence of P^+ and P^-.

COROLLARY 7.2.4 Assume that S *is non-empty. Then the pair* (A,B) *is controllable iff there exist a unique* $P^+ \in S$ *and a unique* $P^- \in S$ *such that* $Re(\lambda(A-BR^{-1}B'P^+)) \leqslant 0$, $Re(\lambda(A-BR^{-1}B'P^-)) \geqslant 0$ *and* $P^- \leqslant P \leqslant P^+$ *for any element* P *of* S.

PROOF Sufficiency follows from Theorems 7.2.1 and 7.2.3, since (A,B) controllable implies (A,B) and (-A,B) stabilizable. To show necessity, again from Theorems 7.2.1 and 7.2.3, (A,B) and (-A,B) are stabilizable. But (see (7.1.8))

(A,B) is stabilizable iff $\chi^+(A) \subset \{A|B\}$[1];

(-A,B) is stabilizable iff $\chi^+(-A) \subset \{-A|B\} = \{A|B\}$.

Hence, $\chi^+(A) + \chi^-(A) \subset \{A|B\}$, therefore in view of (7.1.7) we conclude that

$$\{A|B\} = \mathbb{R}^n .$$ □

An interesting and important situation arises when there is a positive definite gap, namely $P^+ - P^- > 0$. We shall refer to this as a 'positive

---

1) We denote by $\{A|B\} \triangleq \text{Range}\,([\,B \vdots AB \vdots \text{----} \vdots A^{n-1}B\,])$.

gap'. In this connection we have the following.

THEOREM 7.2.5 [5] *Assume that one of the pairs* (A,B) *or* $(-A,B)$ *is stabilizable. Then* $P_1, P_2 \in S$ *and* $P_2 < P_1$ *imply that* $P_1 = P^+$, $P_2 = P^-$, $Re(\lambda(A-BR^{-1}B'P_1)) < 0$, $Re(\lambda(A-BR^{-1}B'P_2)) > 0$ *and the pair* (A,B) *is controllable.* □

Several equivalent statements of the positive gap assumption are given in the following lemma.

LEMMA 7.2.6 [5] *Assume* S *not empty. Then any two of the following conditions imply the remaining one.*

1. *The* ARE *has a positive gap.*
2. $Re(\lambda(A_+)) < 0$ for $A_+ = A - BR^{-1}B'P^+$, $P^+ \in S$.
3. (A,B) *is a controllable pair.* □

Having discussed the ordering of solutions with respect to the special solutions P^+ and P^-, if they exist, we turn to a study of the number of solutions to the ARE and discover more properties of the solutions. We shall first require the following notation. Define $A_i \triangleq A-BR^{-1}B'P_i$ and write the difference $\Delta \triangleq P_i - P$, P_i, $P \in S$. It is readily verifiable that Δ must satisfy a simplified ARE

$$A_i' \Delta + \Delta A_i + \Delta BR^{-1}B'\Delta = 0. \qquad (7.2.2)$$

We now give an interesting and useful lemma with a simple proof.

LEMMA 7.2.7 *Let* Δ *solve the simplified* ARE (7.2.2). *Then* $\ker(\Delta)$ *is an invariant subspace under* A_i.

PROOF Let $x \in Ker(\Delta)$; postmultiplication of (7.2.2) by x gives $A_i' \Delta x + \Delta A_i x + \Delta BR^{-1}B'\Delta x = 0$, or $\Delta A_i x = 0$, which shows $A_i x \in Ker(\Delta)$, i.e. the subspace $Ker(\Delta)$ is invariant under A_i. □

Molinari [7], generalizing the ideas of Kucera [8], originally considered these invariant subspaces and made the simple observation of Lemma 7.2.7, which shows that there is an A_i-invariant subspace associated with each solution Δ of (7.2.2).

The following, however, is not as easily proved: there is a solution Δ associated with certain A_i-invariant subspaces, i.e. there is a one-to-one correspondence between certain A_i-invariant subspaces (to be made more precise shortly) and solutions Δ of (7.2.2). In order to establish this important relation we define the sets

$$\mathcal{S} \triangleq \{\nu | \nu \supset \chi^0(A_i),\ A_i\ \nu \subset \nu, \{A_i|B\} \oplus \nu = R^n\}$$

$$\mathcal{D} \triangleq \{\Delta | \Delta = \Delta' \in R^{n\times n}, \Delta \text{ solves (7.2.2)(and } \mathrm{Re}(\lambda(A_i)) \leqslant 0).\}$$

Next we define a map $M:\mathcal{D} \to \mathcal{S}$ by the formula $\nu = \ker(\Delta)$. The claimed one-to-one correspondence between the elements $\Delta \in \mathcal{D}$ and the elements $\nu \in S$ will now be established by a series of lemmas.

LEMMA 7.2.8 [5] *Let* Δ_1 *and* Δ_2 *solve* (7.2.2) *with* (A_i,B) *stabilizable and* $\mathrm{Re}(\lambda(A_i)) \leqslant 0$. *Then* $\Delta_1 \neq \Delta_2$ *iff* $\mathrm{Ker}(\Delta_1) \neq \mathrm{Ker}(\Delta_2)$ *and* $\Delta_1 = \Delta_2$ *iff* $\mathrm{Ker}(\Delta_1) = \mathrm{Ker}(\Delta_2)$. □

This lemma ascertains that the map M is one-to-one; indeed, let $\nu_1, \nu_2 \in \mathcal{S}$ be the images of $\Delta_1, \Delta_2 \in \mathcal{D}$ under M. Clearly $\Delta_1 = \Delta_2$ im= plies $\nu_1 = \nu_2$. If $\nu_1 = \nu_2$, we must have $\ker(\Delta_1) = \ker(\Delta_2)$, but by the previous lemma 7.2.8, then $\Delta_1 = \Delta_2$. □

Next we shall require the following facts from linear algebra [4].

LEMMA 7.2.9 [5] (a) *Let* (A,B) *be* n×n *and* n×m *matrices respectively, and let* ν *be a k-dimensional subspace of* R^n *which is invariant under* A. *Upon choosing a basis of* R^n *such that the first vectors in the basis span the subspace* ν, *we have that the subspace* ν *determines an* $(n-k) \times (n-k)$ *matrix* A_{22} *and an* $(n-k) \times m$ *matrix* B_2 *such that in the new basis*

$$A \rightarrow \begin{bmatrix} A_{11} & A_{12} \\ 0 & A_{22} \end{bmatrix}, \quad B \rightarrow \begin{bmatrix} B_1 \\ B_2 \end{bmatrix}$$; *moreover the two properties* $Re(\lambda(A_{22})) < 0$ *and controllability of the pair* (A_{22}, B_2) *are invariant with respect to the choice of the basis vectors.*

(b) *Let the subspace* ν *be invariant under* A. *Choose a basis of* R^n *such that the first vectors in the basis span the subspace* ν, *so that in the new basis (see part (a)).*

$$A \rightarrow \begin{bmatrix} A_{11} & A_{12} \\ 0 & A_{22} \end{bmatrix}, \quad B \rightarrow \begin{bmatrix} B_1 \\ B_2 \end{bmatrix}.$$

Then the system (A,B) *is controllable to the subspace target* ν *iff the pair* (A_{22}, B_2) *is controllable, which is equivalent to:* $\{A|B\} + \nu = \mathbb{R}^n$. □

Finally, the following lemma is proved in [5].

LEMMA 7.2.10 *The map* M *is surjective.* □

We are now able to state the following theorem.

THEOREM 7.2.11 [5] *Assume* (A,B) *is stabilizable and* S *is non-empty. From Theorem 7.2.1 we know that there is a maximal solution* $P^+ \in S$. *Let* $A_+ \triangleq A - BR^{-1}B'P^+$. *Then the number of solutions to the ARE is equal to the number of subspaces* M *of* $\mathbb{R}^n$ *such that* $A_+M \subset M, M \supset \chi^0(A_+)$, *and* $\{A|B\} + M = \mathbb{R}^n$. □

Although it would be laborious to write them all down, an explicit count of the number of solutions is now readily derivable. Related results for the case $Q \geqslant 0$ as well as ordering results are given by Kucera [8].

COROLLARY 7.2.12 [5] *Assume (A,B) stabilizable and let* $P^+ \in S$ *be the maximal solution. Then there exists a transformation* T, *det* $T \neq 0$, *such that*

$$T^{-1}A_+T = T^{-1}(A-BR^{-1}B'P^+)T = \begin{pmatrix} A_c & A_{cu} \\ 0 & Au \end{pmatrix},$$

$$T^{-1}B = \begin{pmatrix} B_c \\ 0 \end{pmatrix}, \quad (A_c, B_c) \text{ is controllable}$$

and

$$Re(\lambda(Au)) < 0.$$

Let $\{\lambda_i\}$ be the set of eigenvalues for A_c.

Then the number of real symmetric solutions to the ARE is

(a) infinite iff A_c is not cyclic
(i.e. there is more than one block in the Jordan canonical form of A_c which corresponds to an eigenvalue λ);

(b) finite otherwise, and given by

$$\prod_{\substack{\text{all chains of } A_c \\ Im(\lambda_i) \geq 0 \\ Re(\lambda_i) \neq 0}} (\ell_{chain} + 1)$$

where ℓ_{chain} is the length of a chain in the Jordan form for A_c. □

Several examples of the various situations covered in Corollary 7.2.12 are given by Bucy [9]. Moreover, from the previous corollary it can be seen that if (A,B) is controllable, the ARE has a finite number of real symmetric solutions iff no two Jordan blocks of A_+ have equal negative eigenvalues; in particular, if $Re(\lambda(A_+)) < 0$, then the number of solu= tions to the ARE equals the number of A_+ invariant subspaces of $\mathbb{R}^n$.

It follows that in this case the number of elements in S is finite iff A_+ is cyclic. Moreover, if (A,B) is controllable and S is not empty, then $P = P^+$ is the unique solution to the ARE iff $\chi^0\ (A-BR^{-1}B'P) = \mathbb{R}^n$. These remarks are illustrated in the following two examples.

EXAMPLE 7.2.13 Let $A = 0$, $B = \begin{bmatrix} 1 & 0 \\ 0 & 1 \end{bmatrix}$, $Q = \begin{bmatrix} 1 & 0 \\ 0 & 1 \end{bmatrix}$, $R = I$. Then we have the 'quadratic' equation $P^2 = I$, and $P^+ = I$ (while $P^- = -I$) so that $A_+ = A-BR^{-1}B'P^+ = -I = \begin{bmatrix} -1 & 0 \\ 0 & -1 \end{bmatrix}$.

Thus, since we here have two blocks in the Jordan canonical form of A_+ which correspond to the eigenvalue -1 (i.e. A_+ is not cyclic) we con= clude from Corollary 11 that the ARE has an infinite number of solutions, which are given by

$$P(\theta) = \begin{bmatrix} \cos\theta & \sin\theta \\ +\sin\theta & -\cos\theta \end{bmatrix}, \ 0 \leqslant \theta < 2\pi.$$

EXAMPLE 7.2.14 Consider the scalar controllable problem ($b \neq 0$). The Riccati equation is $p^2b^2 - 2ap - q = 0$, with solutions $(a \pm \sqrt{a^2+qb^2})/b^2$. Then $\chi^0(A_+) = \mathbb{R}^n$ is equivalent to $0 = a-b^2(a \pm \sqrt{a+qb^2})/b^2 = \sqrt{a^2+qb^2}$. Hence, we see that the quadratic equation has a unique solution iff the discriminant $a^2+qb^2 = 0$.

As shown in Theorem 7.2.1, assuming stabilizability of (A,B), all solu= tions may be ordered with respect to a maximal solution P^+. Further= more, we know that all solutions are associated with certain A_+ invariant subspaces. The last theorem in this section tells us that the ordering of the solutions to the ARE has a one-to-one correspondence with an ordering of the related subspaces ν.

THEOREM 7.2.15 [5] *Assume* $P_1, P_2 \in S$, $\Delta_1 = P^+ - P_1$, $\Delta_2 = P^+ - P_2$, (A,B) *stabilizable*, $\nu_1 = \ker(\Delta_1)$, $\nu_2 = \ker(\Delta_2)$. *Then* $P_1 \geqslant P_2$ *iff* $\nu_1 \supset \nu_2$.□

7.3 RELATIONSHIP TO POTTER'S METHOD

An older approach to generating solutions to the ARE was given by Potter [3], and later generalized by others (Bass [10], Mårtenson [11]). We shall in this section review Potter's method and compare it with the results of Section 7.2. The Potter-Bass spectrum-factorization method (in the state space) is important because at present it is the most practical numerical method for computing P^+, although for obtaining theoretical results it is not as attractive as the approach of Section 7.2. The basis for the idea is summarized in the theorem below. For further details, see the references [3,10,11]. First, we define the *Hamiltonian matrix* $\mathcal{H} \in \mathbb{R}^{2n\times 2n}$

$$\mathcal{H} \triangleq \begin{bmatrix} A & -BR^{-1}B' \\ -Q & -A' \end{bmatrix}.$$

THEOREM 7.3.1 (Potter, Mårtenson) Let each eigenvector or generalized eigenvector a_i *of* $\mathcal{H}$ *be partitioned into two* $n\times 1$ *vectors* $a_i' = [x_i' \,\vdots\, y_i']$. *Then every solution of the* ARE *takes the form* $P = YX^{-1}$ *where the* $n\times n$ *matrices* $X = [x_1, x_2, \ldots, x_n]$, $Y = [y_1, y_2, \ldots, y_n]$ *are so chosen that* $\det(X) \neq 0$, *and if any generalized eigenvector is chosen, all of the lower-ranking eigenvectors are also included. In order that the solutions generated by the above scheme may be real and symmetric it is necessary and sufficient that* $\lambda_i + \lambda_j \neq 0$ *for all* i,j *in the included set used to form* X *and* Y, *and if* a_i *is complex, its conjugate must also be included.* □

The connection with Section 7.2 is brought out by the following two lemmas.

LEMMA 7.3.2 [5] If $P_i \in S$, $A_i \triangleq A-BR^{-1}B'P_i$, *then* $\mathcal{H}$ *is similar* to

$$\begin{bmatrix} A_i & -BR^{-1}B' \\ 0 & -A_i' \end{bmatrix}.$$

Moreover, if in addition $\chi^-(A_+) \supset \{A|B\}$ *, then* $\mathcal{H}$ *is similar to the block-diagonal matrix* $\text{diag}(A_+, -A_+')$. □

Note that the additional hypothesis of Lemma 7.3.2 is actually $Re(\lambda(A_+)) < 0$ if (A,B) is stabilizable(i.e. it is the 'positive gap' assumption of Section 7.2), since then (A_+,B) is stabilizable and $\chi^-(A_+) \supset \{A|B\} \supset \chi^+(A_+)$ yields $\chi^+(A_+) = 0$. In this case Potter's method is simplified, as shown below.

LEMMA 7.3.3 [5] *Assume* $Re(\lambda(A_+)) < 0$. *Then the eigenvectors of* $\mathcal{H} = \begin{bmatrix} A_+ & -BR^{-1}B' \\ 0 & -A_+' \end{bmatrix}$ *are given by the columns of the* $2n\times 2n$ *matrix* $\begin{bmatrix} M_x & -SM_y \\ 0 & M_y \end{bmatrix}$ *where* $A_+M_x = M_x\Lambda_1$, $-A_+'M_y = M_y\Lambda_2$, *i.e.* M_x *and* M_y *are modal matrices for* A_+ *and* $-A_+'$ *and* Λ_1 *and* Λ_2 *on the corresponding Jordan forms* ; *the* $n\times n$ *matrix* $S = S'$ *is the unique positive semi-definite solution to*

$$A_+S + SA_+ - BR^{-1}B' = 0.$$ □

Note that in the case $Re(\lambda(A_+)) < 0$, (A,B) controllable, the method of the previous section reduces to considering all (real) A_+ invariant subspaces. This is of course equivalent to computing the modal matrix for A_+ as required in the previous lemma; selection of the vectors a_i to be included in X and Y when the method is applied, is equivalent to selection of an invariant subspace ν. The maximal and minimal solutions P^+ and P^- can be interpreted as special choices of the a_i vectors in solving the Δ equation (7.2.2). If we pick $X = M_x$, $Y = 0$, we get $\Delta = YX^{-1} = 0$ corresponding to P^+. On the other hand, if we select $X = -SM_Y$, $Y = M_Y$, then $\Delta = M_Y(-SM_Y)^{-1} = -S^{-1}$ (if det $S \neq 0$), namely $P^- = P^+ - S^{-1}$. In view of Lemma 7.3.2 we can also easily conclude that if there exist solutions to the ARE then the Hamiltonian matrix $\mathcal{H}$ has no imaginary eigenvalues of odd multiplicity.

EXAMPLE 7.3.4 In the scalar case we have the ARE : $2pa - b^2p^2 + q = 0$. We note that the condition for real solutions is that the discriminant $a^2 - h^2q \geqslant 0$, which is the same as the condition of the theorem, namely that

$$\mathcal{H} = \begin{bmatrix} a & -b^2 \\ -q & -a \end{bmatrix}$$

has no imaginary eigenvalues. For n = 2, it is easy to show that the corresponding condition is then $\det(\mathcal{H}) \geqslant 0$.

EXAMPLE 7.3.5 Assume that the pair (A,B) is controllable and n = 2, m = 1. Without loss of generality we assume that

$$A = \begin{bmatrix} 0 & 1 \\ a_1 & a_2 \end{bmatrix}, \; b = \begin{bmatrix} 0 \\ 1 \end{bmatrix}, \quad Q = \begin{bmatrix} q_1 & q_2 \\ q_2 & q_3 \end{bmatrix}.$$

Then the ARE has a unique solution iff $a_1^2 + q_1 \geqslant 0$ and $a_2^2 + q_3 = -2(a_1+\sqrt{a_1^2+q_1})$.

PROOF $\mathcal{H} = \begin{bmatrix} A & -BB' \\ -Q & -A' \end{bmatrix} = \begin{bmatrix} 0 & 1 & 0 & 0 \\ a_1 & a_2 & 0 & -1 \\ -q_1 & -q_2 & 0 & -a_1 \\ -q_2 & -q_3 & -1 & -a_2 \end{bmatrix}$. Upon requiring that

the eigenvalues of $\mathcal{H}$ are pure imaginary and repeated we obtain:

$a_1^2 + q_1 \geqslant 0$, $a_2^2 + q_3 = -2(a_1 + \sqrt{a_1^2 + q_1})$. Note that the above condition corresponds to the condition that in the quadratic equation (i.e. when n = 1) the discriminant must be zero. □

7.4 ADDITIONAL PROPERTIES OF SOLUTIONS

Although we now have a good deal of information concerning the solutions to the ARE, the significance of these solutions in relation to the re= gulator problem has yet to be pointed out. In order to do this, several additional properties of the solutions will be developed; in particular the properties of the closed-loop system matrix $A-BR^{-1}B'P$, $P \in S$.

Our first result in this respect is Theorem 7.4.1, which shows that $A-BR^{-1}B'P$ invariant subspaces associated with eigenvalues on the imaginary axis are the same for all $P \in S$, and that the degree of in= stability of the closed-loop system increases as we depart from the solution P^+.

THEOREM 7.4.1 [5] *Assume* $P_1, P_2 \in S$, (A,B) *stabilizable and let* $P_i \triangleq P_+ - \Delta_i$, i=1,2. *Then,*

(a) *if* $P_1 \geqslant P_2$ *then* $\chi^+(A-BR^{-1}B'P_2) \supset \chi^+(A-BR^{-1}B'P_1)$;

(b) $\chi^0(A-BR^{-1}B'P_1) = \chi^0(A-BR^{-1}B'P_2)$. □

Part (b) of this Theorem originally appeared as remark 19 in Willems's paper [6], but without a proof. Since all closed-loop system matrices $A-BR^{-1}B'P$ thus contain the same invariant subspace $\chi^0(A_+)$, two special cases are of interest: $\chi^0(A_+) = 0$ and $\chi^0(A_+) = \mathbb{R}^n$. The first case corresponds to the positive gap situation and the second case leads to a unique solution to the ARE. Indeed, since $\nu = \ker(\Delta) \supset \chi^0(A_+) = R^n$, we have a unique solution $\Delta = 0$ to the ARE, and hence $P = P^+$ is the only element in S.

It is well known (see also Lecture 4) that the solution to the linear regulator problem is given in terms of the solution to the ARE. In fact, if P satisfies the ARE (7.1.5), then (7.1.2) is minimized with respect to $u(\cdot)$, subject to (7.1.1) and $T \to \infty$, by the choice

$$u(t) = -R^{-1}B'Px(t). \qquad (7.4.1)$$

If (7.1.5) and (7.4.1) hold, we find

$$J = x_0' Px_0 - x'(\infty)Px(\infty).$$

Thus, if P is stabilizing, i.e. $Re(\lambda(A-BR^{-1}B'P)) < 0$, then $x(\infty) = 0$ and, if more than one P is stabilizing, we so choose P as to minimize $x_0' Px_0$ However, from Theorem 7.2.1 only one solution P has $\mathrm{Re}(\lambda(A-BR^{-1}B'P)) \leqslant 0$ in general, so that $P = P^+$ solves the regulator problem. However, if we do not require $x(\infty)$ to be bounded, J will be bounded if

$$\chi^+(A-BR^{-1}B'P) \subset \mathrm{Ker}(P).$$

In fact, the following lemma holds.

LEMMA 7.4.2 [5] *Let* $P_1, P_2 \in S$. *Then if* $\mathrm{Ker}(P_1) = \mathrm{Ker}(P_2)$ *and* $\chi^+(A-BR^{-1}B'P_1) \subset \mathrm{Ker}(P_1)$, $\chi^+(A-BR^{-1}B'P_2) \subset \mathrm{Ker}(P_2)$, *this implies that* $P_1 = P_2$. □

We also have the following lemma which orders the kernels of the solu= tions to the ARE.

LEMMA 7.4.3 [5] *Let* $P \in S$, *with* $Q \geqslant 0$. *Then* $\ker(P) \subset (\ker(Q))_A$. □

We now have the following ordering result for positive semi-definite solutions of the ARE.

THEOREM 7.4.4 *Assume* $Q \geqslant 0$. *Let* $P \in S$, $P \geqslant 0$ *and* $\ker(P) = (\ker(Q))_A$. *Then* $P \leqslant K$ *for any* $K \in S$, $K \geqslant 0$. □

In fact this P yields the solution to the regulator problem with $Q_f = 0$.

7.5 THE DIFFERENTIAL MATRIX RICCATI EQUATION

The behaviour of the solutions to the RE and the regions of attraction for the solutions to the ARE are important in linear-quadratic regulator theory, as shown in the simple scalar example with terminal payoff:

$$J = x^2(T)\, Q_f + \int_0^T u^2(t)dt, \; T \to \infty$$

and dynamics

$$\dot{x} = ax + bu, \; x(0) = x_0.$$

Assuming $b \neq 0$, the system is controllable, so that the ARE has maximal and minimal solutions P^+ and P^-. The solutions are 0 and $2a/b^2$, and since $P^+ \geqslant P^-$, if $a < 0$ we have $P^+ = 0$ and $P^- = 2a/b^2$; otherwise $P^+ = 2a/b^2$, $P^- = 0$. Therefore, denoting by P the solution of the ARE which determines the solution to the regulator problem, if $Q_f = 0$ then $P = 0$ for all a. If $Q_f > 0$, then P switches to 0 if $a < 0$, and $P = 2a/b^2$ if $a \geqslant 0$. If $Q_f < 0$, then if $a \geqslant 0$ or if $a < o$ and $Q_f < 2a/b^2$ there is no solution; if $a < o$ and $Q_f = 2a/b^2$, then $P = 2a/b^2$, and if $Q_f > 2a/b^2$, then $P = 0$.

We begin by giving a simple bound on the solution of the RE.

THEOREM 7.5.1 [5] *Let* $P_1, P_2 \in S$ *and* $P(0) = P'(0)$ *satisfy* $P_1 \leqslant P(0) \leqslant P_2$. *Then there exists a solution* $P(t)$ *to the* RE *on* $[0,\infty)$ *and* $P_1 \leqslant P(t) \leqslant P_2$.□

Note however that two solutions $P_1, P_2 \in S$ can be ordered iff $\nu_1 \subset \nu_2$ or $\nu_2 \subset \nu_1$ with ν_1 and ν_2 defined as in Theorem 7.2.15.

Next, we present the following important result.

THEOREM 7.5.2 [5] *Assume that the pair*(A,B) *is controllable, that* S *is non-empty, that we have a positive gap, and that* $P(0) > P^-$ *(namely* $Q_f = P(0) > P^-$ *Then* $P(t)$ *exists on* $[0,\infty)$, $P(t) > P^-$ *and* $\lim_{t \to \infty} P(t) = P^+$. □

On the other hand, other initial conditions lead to unbounded solutions, as shown in the following corollary.

COROLLARY 7.5.3 [5] *Assume* $P^+ > P^-$ *and consider the* RE *with initial condition* $P(0)$, $P(0) \not\geqslant P_i$, *for any* $P_i \in S$, $\det(P(0) - P^-) \neq 0$. *Then for some* $t^* \in [0,\infty)$, *the solution of* $P(t)$ *becomes unbounded.* □

Furthermore, a solution to the RE may exist on $[0,\infty)$ even if the hypo= thesis of Theorem 7.5.1 does not hold. In this respect consider the following example.

EXAMPLE 7.5.4 Let $A = \begin{bmatrix} 0 & 1 \\ 0 & 0 \end{bmatrix}$, $B = \begin{bmatrix} 0 \\ 1 \end{bmatrix}$, $Q = \begin{bmatrix} 0 & 0 \\ 0 & 1 \end{bmatrix}$, $P(0) = \begin{bmatrix} 0 & 0 \\ 0 & 2 \end{bmatrix}$.

Then $P^- = \begin{bmatrix} 0 & 0 \\ 0 & -1 \end{bmatrix}$, $P^+ = \begin{bmatrix} 0 & 0 \\ 0 & 1 \end{bmatrix}$. In this case $P(0) - P^- = \begin{pmatrix} 0 & 0 \\ 0 & 3 \end{pmatrix} \not> 0$, (as in the Theorem hypothesis) yet it turns out that P(t) exists on $[0,\infty)$.

In the paper by Rodriguez-Canabal [12] it is assumed that

A_1. S_+ is bounded above and below, where

$$S_+ \triangleq \{P | P = P', PA + A'P - PBR^{-1}B'P + Q \geqslant 0\};$$

A_2. $P^+ - P^-$ is nonsingular.

The assumption A_1 is equivalent to the existence of $P^+ \geqslant P^-$, and A_2 is the positive gap assumption when combined with A_1. In Theorem 7.5.2 and in Corollary 7.5.3 the same assumptions are made, and in fact Theorem 3.1 of Rodriguez-Canabal contains the same result. It is, how= ever, possible to ascertain existence of the solution to the ARE without requiring the assumption A_2; we refer to the above example.

7.6 SOME CONNECTIONS WITH DIFFERENTIAL GAMES

An interesting type of Riccati equation occurs in the study of the linear-quadratic differential game

$$\dot{x} = Ax + Bu + Cv,$$

$$y = Hx$$

where the two control variables $u \in R^p$ and $v \in R^q$ are the controls of the minimizer and maximizer respectively. The corresponding Riccati equation (GARE) is

$$A'P + PA - P(BB' - CC')P + H'H = 0. \qquad (7.6.1)$$

Note that now the term BB' - CC' is indefinite. In this section we shall apply to this equation some of the results obtained for the ARE. Specifically, from Theorem 7.2.3 we obtain the following ordering result.

THEOREM 7.6.1 [5] *Assume* (A,H) *detectable (see Section* 7.1.1), *and that there exists a real symmetric negative definite solution* K *to the* GARE (7.6.1). *Then there exists a real symmetric negative definite solution* $\bar{K}$ *such that* $K \leq \bar{K}$ *for* K *any other non-singular solution to the* GARE *and* $Re(\lambda(A-(BB' - CC')\bar{K}) \geq 0$. □

Analogously, if (-A,H) is detectable, and there exists a real symmetric positive-definite solution K to the GARE, then there exists a real symmetric positive-definite solution $\bar{\bar{K}}$ such that $K \geq \bar{\bar{K}}$ for K any other non-singular solution to the GARE, and $Re(\lambda(A-(BB' - CC')\bar{\bar{K}}) \leq 0$. Hence, if the pair (A,H) is observable, it cannot be that the GARE has both real symmetric positive-definite and negative definite solutions simultaneously. Furthermore, in view of the above ordering result, Proposition 1 of Mageirou [13] and Theorem 1 in [14], it is possible (if the ARE has a positive-definite solution) to compute the P^+ solution to the optimal control ARE as follows.

THEOREM 7.6.2 *Consider the* ARE *and assume that the pair* (A,B) *is controllable. If the* ARE *has a real symmetric positive definite solution, then it is possible to find* P^+:

$$P^+ = [\lim_{t \to \infty} K(t)]^{-1},$$

where K(t) *is the monotonically increasing solution of*

$$\dot{K} = -AK - KA' - KQK + BB', \; K(0) = 0.$$

Note that this integration procedure is stable. □

We also have the following further connections between the ARE (7.1.5) and the GARE:

$$0 = AK + KA' + KQK - BB' \qquad (7.6.2)$$

Equation (7.1.5) was our object of study in Sections 7.2-7.5, whereas equation (7.6.2) would arise in infinite-time linear-quadratic differen= tial games which correspond to the system $(-A', Q = G_1G_1' - G_2G_2', B')$. On the other hand, one obtains equation (7.6.2) from equation 7.1.5) upon setting $K = P^{-1}$ so that, obviously, if there exists a real symmetric non-singular solution to equation (7.1.5), then there exists a real symmetric non-singular solution to equation (7.6.2), and vice versa.

We shall now further examine the properties of the solutions of equations (7.1.5) and (7.6.2).

Choose a basis for $\mathbb{R}^n$ such that the first vectors in the basis span $(\mathrm{Ker}(B'))_{(-A')} = \{A|B\}^{\perp}$. In this basis

$$-A' = \begin{bmatrix} -A_{11}' & -A_{21}' \\ 0 & -A_{22}' \end{bmatrix}, \quad B' = (0, B_2'), \quad Q = \begin{bmatrix} Q_{11} & Q_{12} \\ Q_{12} & Q_{22} \end{bmatrix},$$

$K = \begin{bmatrix} K_{11} & K_{12} \\ K_{12} & K_{22} \end{bmatrix}$, and the pair $(-A_{22}', B_2)$ is observable iff the pair (A_{22}, B_2) is controllable. Consider the corresponding reduced order algebraic Riccati equation

$$0 = A_{22} K_{22} + K_{22} A_{22}' + K_{22} Q_{22} K_{22} - B_2B_2' \qquad (7.6.3)$$

and the conditions

(i) there exists a real symmetric positive semi-definite solution to (7.6.2);

(ii) there exists a real symmetric positive definite solution to (7.6.3);

(iii) there exists a real symmetric solution to (7.1.5);

(iv) there exists a real symmetric non-singular solution to (7.1.5);

(v) there exists a real symmetric non-singular solution to (7.6.2).

Then the following hold:

1. (i) is equivalent to (ii);
2. (ii) plus (A,B) stabilizable implies (iii);
3. (iii) plus Q non-singular implies (iv);
4. (iv) is equivalent to (v).

Hence we also have the following corollary.

COROLLARY 7.6.3 Assume that the pair (A,B) *is stabilizable. If there exists a real symmetric positive semi-definite solution to* (7.6.2), *then there exists a real symmetric solution to* (7.1.5). □

In fact, (i) is a standing assumption in Mageirou [13].

7.7 CONCLUDING REMARKS

We relaxed the standard controllability assumption of the pair (A,B) to stabilizability of the pair (A,B). We then characterized the solutions to the ARE in terms of ordering and 'closed-loop stability properties'; moreover it was shown that stabilizability is really a necessary condi= tion for the existence of P^+ (and similarly, stabilizability of (-A,B) is necessary for the existence of P^-).

We then turned our attention to the question of an infinite/finite number of solutions to the ARE. In the latter case we gave a count of the number of solutions to the ARE. Also, it was noted that the con= dition for a unique real symmetric solution, or exactly two such solu= tions to the ARE, is a nice generalization of the discriminant condition for the quadratic equation. In addition, we discussed the relation of our approach to the Potter-Bass method.

Obviously, the solutions of the ARE are equilibrium points in the RE. Motivated by the fact that in the linear-quadratic regulator problem in the terminal payoff, it is rather the behaviour of the solution to the RE which determines the optimal control and payoff, we then investi= gated the existence of solutions to the RE and also considered the regions of attraction in the RE.

Our results and techniques were also brought to bear on the analysis of the Riccati equation which corresponds to the simple zero-sum linear quadratic infinite-time differential game (the indefinite R case). We have not, however, discussed the singular (R positive semidefinite) case and instead refer the reader to reference [15].

7.8 REFERENCES

[1] Kalman, R.E., Contributions to the theory of optimal control. *Bol. Soc. Mat. Mex.*, Vol 5, 1960, pp 102-199.

[2] Wonham, W.M., On a matrix Riccati equation of stochastic control. *SIAM J. Control*, Vol 6, 1968, pp 681-698.

[3] Potter, J.E., Matrix quadratic solutions. *SIAM J. on Appl. Math.*, Vol 11, No 3, May 1966, pp 496-501.

[4] Gantmacher, F.R., *The Theory of Matrices*. Chelsea, N.Y., 1960.

[5] Pachter, M., Bullock, T.E., Ordering and stability properties of the Riccati equation, *NRIMS Technical Report*, WISK 264, CSIR, Pretoria, 1977.

[6] Willems, J.C., Least squares stationary optimal control and the algebraic Riccati equation. *IEEE Trans. Automat. Contr.*, Vol AC-16 Dec 1971, pp 621-634.

[7] Molinari, B.P., The time-invariant linear-quadratic optimal control problem. *Automatica,* Vol 13, pp 347-357, 1977.

[8] Kucera, V., On non-negative definite solutions to matrix quadratic equations. *Automatica*, Vol 8, 1972, pp 413-423.

[9] Bucy, R.S., Structural stability for the Riccati equations. *SIAM J. Control*, Vol 13, no 4, July 1975, pp 749-753.

[10] Bass, R.W., Machine solution of high-order matrix Riccati equations. *Douglas Aircraft Report, Missiles & Space Systems Division,* Santa Monica, 1967.

[11] Mårtenson, K., On the matrix Riccati equation. *Inform. Sci.* Vol 3, 1971, pp 17-49.

[12] Rodriguez-Canabal, T., The geometry of the Riccati equation. *Stochastics* Vol 1, 1973, pp 129-149.

[13] Mageirou ,E.F., Values and strategies for infinite-time linear-quadratic games. *IEEE Trans. Automatic Control* Vol AC-21, 1976, pp 547-550.

[14] Pachter, M., Some properties of the value and strategies in infinite-time linear-quadratic differential games. *IEEE Trans. Automatic Control* Vol AC-23, 1978, pp 746-748.

[15] Jacobson, D.H., Totally singular quadratic minimization problems. *IEEE Trans. Automatic Control* Vol AC-16, 1971, pp 651-658.

ADDITIONAL BIBLIOGRAPHY

The following paper is complementary to [6]

[16] Willems, J.C., On the existence of a non-positive solution to the Riccati equation. *IEEE Trans. Automatic Control* Vol AC-19, Oct 1974, pp 592-593.

In addition, we recommend:

[17] Kucera,V., A contribution to matrix quadratic equations. *IEEE Trans. Automatic Control* Vol AC-17, June 1972, pp 344-347.

[18] Molinari, B.P., The stabilizing solution of the algebraic Riccati equation. *SIAM J. Contr.* Vol 11, No 2, May 1973, pp 262-271.

[19] Simaan , M., A note on the stabilizing solution of the algebraic Riccati equation. *Int. J. Cont.* Vol 20, no 2, 1974, pp 239-241.

[20] Bucy, R.S., Joseph, P.D., *Filtering for Stochastic Processes, with Applications to Guidance.* Interscience, New York, 1968.

[21] Brockett, R.W., *Structural Properties of the Equilibrium Solutions of Riccati Equations.* Springer-Verlag, Lecture Notes on Mathema= tics, Vol 132, 1969, pp 61-69.

[22] Bucy, R.S., The Riccati equation and its bounds. *Journal of Computer and System Sciences*, Vol 6, 1972, pp 343-353.

[23] Casti, J., The linear quadratic control problem : Some recent results and outstanding problems. Forthcoming survey article.

8. CONTROLLABILITY SUBJECT TO CONTROLLER CONSTRAINTS

8.1 INTRODUCTION

In the present lecture we shall consider the controllability and reach= ability problem associated with the control system

$$\frac{dx}{dt} = Ax + Bu,\ x(0) = x_o,\quad t \geqslant 0; \qquad (8.1.1)$$

here the state vectors x, $x_o \in \mathbb{R}^n$, the control vector $u \in \mathbb{R}^m$ and A and B are $n\times n$ and $n\times m$ matrices, respectively. In addition, we suppose that a fixed subset $\Omega \subseteq \mathbb{R}^m$ is given such that admissible controllers $u(\cdot)$ are those which are piecewise continuous and satisfy $u(t) \in \Omega$ for all $t \in [0,t_1]$.

The following theorem is well known (see for example, [1] and [2]).

THEOREM 8.1.1 [1] *Consider the control system* (8.1.1). *Assume that* Ω *is compact.* (a) *Then the reachable set from* $x_o = 0$:

$$R(t_1) \triangleq \{ \int_0^{t_1} e^{A(t_1-\tau)}Bu(\tau)d\tau \,|\, u(\tau) \in \Omega\} \quad , t_1 > 0,$$

is compact and convex; moreover if Ω *is relaxed to its convex hull and if* $R_{\text{convexhull}(\Omega)}(t_1)$ *is the corresponding reachable set for all controllers* $u(t) \in \text{convexhull}(\Omega)$ *on* $0 \leqslant t \leqslant t_1$, *then* $R(t_1) = R_{\text{convexhull}(\Omega)}(t_1)$.

(b) *If, in addition,* $0 \in \text{interior}(\Omega)$ *then a necessary and sufficient condition for null-controllability (and reachability), viz.*

$$0 \in \text{interior}(R(t_1))$$

is that

$$\text{range}([B \,|\, AB \,|\, \cdots \,|\, A^{n-1}B]) = \mathbb{R}^n. \qquad \square$$

It is our purpose in the first part of the present lecture to consider the reachability/controllability result of Theorem 8.1.1, with minimal assumptions on the control restraint set Ω, in regard to compactness, convexity and the requirement that $0 \in$ interior (Ω). Thus, in Section 8.2 the compactness and convexity assumptions are relaxed, whereas in Section 8.3 the hypothesis $0 \in$ interior(Ω) is weakened. In Section 8.4 and 8.5 we discuss the concept of arbitrary-interval null-controllability. Finally, in Section 8.6 we comment on connections between the continuity of the minimum-time function and constrained controllability.

8.2 A GENERAL CONTROL RESTRAINT SET

The compactness of the control restraint set does not enter into the relevant arguments on pp.163,77, and 167 of [1] where a proof of Theorem 8.1.1 is given. We thus realize that the following holds.

THEOREM 8.2.1 *Consider the linear control system* (8.1.1) *with any control restraint set* Ω. *The reachable set* $R(t_1)$ *is convex and dense in* $R_{\text{convexhull}(\Omega)}(t_1)$. □

Since for any convex set $C \subset \mathbb{R}^n$,

$$\text{relative interior (closure } (C)) = \text{relative interior } (C) \tag{8.2.1}$$

we also conclude the following.

THEOREM 8.2.2 *Consider the linear control system* (8.1.1) *with any control restraint set* Ω.

Then

$$\text{relative interior } (R(t_1)) = \text{relative interior } (R_{\text{convexhull}(\Omega)}(t_1)).$$

PROOF In view of Theorem 8.2.1, $R(t_1)$ is dense in $R_{\text{convexhull}(\Omega)}(t_1)$ i.e. $R_{\text{convexhull}(\Omega)}(t_1) \subset \text{Closure}\,(R(t_1))$. Thus $\text{closure}(R_{\text{convexhull}(\Omega)}(t_1)) = \text{closure}(R(t_1))$ so that, since $R(t_1)$ is con= vex (Theorem 8.2.1), and in view of (8.2.1),

$$\text{relative interior}\,(R(t_1)) = \text{relative interior}\,(R_{\text{convexhull}(\Omega)}(t_1)).$$
□

These results can be applied (see, for example, [2]) to show the follow= ing result.

THEOREM 8.2.3 *Assume that* $0 \in \text{convexhull}(B\Omega)$. *Then the linear control system* (8.1.1) *is null-controllable iff the system is null-controllable with the control restraint set* convexhull(Ω). □

We conclude this section with an illustration.

EXAMPLE 8.2.4 In a combinatorial control problem one cannot activate all the control channels simultaneously. Specifically, assume that in the control system

$$\frac{dx}{dt} = Ax + u,$$

$u(t) \in \bigcup_{i=1}^{k} \mathcal{B}_i$ for all $t \geqslant 0$, where $\mathcal{B}_i$, $i=1,2,\ldots,k$ are given subspaces of $\mathbb{R}^n$. Let B be the $n \times n$ matrix such that

$\text{range}(B) = \sum_{i=1}^{k} \mathcal{B}_i$. Then since $\text{convexhull}\,(\bigcup_{i=1}^{k} \mathcal{B}_i) = \text{range}(B)$ we conclude that we have controllability iff the pair (A,B) is controllable.

8.3 BRAMMER'S THEOREM

The condition

$$0 \in \text{interior}(\Omega) \tag{8.3.1}$$

in Theorem 8.1.1 is a sufficient condition for null-controllability pro=

vided that

$$\text{range}([B \mid AB \mid \text{----} \mid A^{n-1}B]) = \mathbb{R}^n. \qquad (8.3.2)$$

It turns out, however, that the condition (8.3.1) is not a necessary condition for null-controllability, and the following stronger result is due to Brammer [3].

THEOREM 8.3.1 (Brammer) *For a given subset* $\Omega \subseteq \mathbb{R}^n$ *with*

$$0 \in \Omega, \qquad (8.3.3)$$

consider the control system

$$\dot{x} = Ax + u \quad , \qquad u(t) \in \Omega. \qquad (8.3.4)$$

This control system is null-controllable iff both of the following conditions are satisfied:

$$\text{the linear hull of } (\{e^{-At}u : t \geqslant 0,\ u \in \Omega\}) = \mathbb{R}^n \qquad (8.3.5)$$

(this is equivalent to the condition (8.3.2), *where the matrix* B *is such that*

range(B) = linear hull of (Ω));

and:

none of the real eigenvectors of A' *is an exterior normal to the set* Ω *at* 0. (8.3.6)

PROOF The necessity part of the theorem follows directly from the definition of R(t) [3]. To prove sufficiency, we follow an argument due to O.Hájek [4], which hinges on the following result proved as Lemma 8.3.2 below: If V is a closed and convex cone in $\mathbb{R}^n$ but is not a linear subspace, and if V is invariant under e^{At} for all $t \geqslant 0$, then V contains an eigenvector of A.

Thus, assume that (8.3.5) and (8.3.6) hold, but that $0 \notin$ interior $R(t)$ for all $t > 0$. Since $R(t_1) \subset R(t_2)$ if $t_1 \leqslant t_2$, there must exist a vector $v \neq 0$ which is an exterior normal at 0 to all $R(t)$, viz.

$$v'x \leqslant 0$$

for all $x \in R(t)$, $t \geqslant 0$, for *all* $u(t) \in \Omega$; this in turn implies that

$$v'e^{-At}u \leqslant 0 \text{ for all } t \geqslant 0 \text{ and all } u \in \Omega \tag{8.3.7}$$

Next, define

$$W = \text{conichull } (\{e^{-At}u \mid u \in \Omega,\ t \geqslant 0\}).$$

Thus, by (8.3.7) we conclude that also

$$v'w \leqslant 0 \text{ for all } w \in W.$$

Moreover, by (8.3.5)

$$W + (-W) = \mathbb{R}^n \tag{8.3.8}$$

and by construction W is invariant under e^{-At} for all $t \geqslant 0$, i.e.

$$e^{-At}W \subset W \text{ for all } t \geqslant 0. \tag{8.3.9}$$

Let V be the polar cone of W, viz.

$$V = \{v \mid v'w \leqslant 0 \text{ for all } w \in W\}.$$

Thus V is a closed convex cone, and owing to (8.3.8) we know that V is not a subspace. In addition, (8.3.9) implies the invariance of V under $e^{-A't}$ for all $t \geqslant 0$. Hence the above-mentioned lemma shows that V con= tains an eigenvector v of A', which contradicts (8.3.6). Hence, our assumption that

$0 \notin$ interior $(R(t))$ for all $t \geqslant 0$ leads to a contradiction, and the theorem is proved.

Finally we prove the lemma referred to above. □

LEMMA 8.3.2 *Let* V *be a closed convex cone in* $\mathbb{R}^n$ *which is not a linear subspace of* $\mathbb{R}^n$. *Then*

$$e^{At}V \subset V \text{ for all } t \geqslant 0$$

implies that V *contains an eigenvector of* A.

PROOF Consider the mapping $\pi: \mathbb{R}^n \setminus \{0\} \to \mathbb{R}^n$ given by

$$\pi(x;t) \triangleq \frac{e^{At}x}{|e^{At}x|},$$

and let

$$S \triangleq \{x | x'x = 1\}$$

be the unit sphere of $\mathbb{R}^n$. Obviously, $\pi(V \cap S) \subseteq V \cap S$.

Thus, applying the Lefchetz fixed-point theorem for the above mapping on the compact contractible topological polyhedron $V \cap S$, we conclude that for each $t \geqslant 0$ there exists a fixed point x_t in $V \cap S$.

Taking a sequence of times $t \downarrow 0$ we select a limit point $\hat{x} \in V \cap S$ of the corresponding sequence of fixed points. There is thus a subsequence, which we denote by $t_k \downarrow 0$, such that $x_k \to \hat{x}$. The fixed-point property implies that

$$\frac{1}{t_k}(e^{At_k} - I)x_{t_k} = \frac{1}{t_k}(|e^{At_k}x_k| - 1)x_{t_k}.$$

Letting $k \to \infty$, it follows that for some λ we have

$$A\hat{x} = \lambda\hat{x},$$

which proves the lemma.

8.4 ARBITRARY-INTERVAL NULL-CONTROLLABILITY

It is well known that when convexhull(Ω) contains the origin (of $\mathbb{R}^m$) in its interior, null controllability of (8.1.1) implies that for each $t > 0$ there exists an open set $W(t)$ in $\mathbb{R}^n$ which contains the origin, the set being such that any $x_0 \in W(t)$ can be steered to 0 in time t. When convexhull (Ω) does not contain the origin in its interior this may not be possible, as the following example illustrates.

EXAMPLE 8.4.1 Let $A = \begin{bmatrix} 0 & 1 \\ -1 & 0 \end{bmatrix}$, $b = \begin{bmatrix} 0 \\ 1 \end{bmatrix}$, $x_0 = \begin{bmatrix} x_{10} \\ x_{20} \end{bmatrix}$ and let $\Omega = [0,1]$.

Then

$$x_1(t) = x_{10}\cos t + x_{20}\sin t + \int_0^t \sin\sigma . u(t-\sigma)d\sigma,$$
$$x_2(t) = x_{10}\sin t + x_{20}\cos t + \int_0^t \cos\sigma . u(t-\sigma)d\sigma.$$

If the system trajectories in $\mathbb{R}^2$ are considered, it can be concluded that each initial state can be steered to the origin, i.e. we have null-controllability; however, it is not possible to steer the initial states $x(0) = \binom{0}{\alpha}$, $\alpha > 0$, to the origin in time t, $t < \frac{\pi}{2}$.

We are thus motivated to introduce the following definition.

DEFINITION *The system* (8.1.1) *is arbitrary-interval null-controllable if for each* $t > 0$ *there exists an open set* $W(t) \subset \mathbb{R}^n$ *which contains the origin and for which* $x_0 \in W(t)$ *can be controlled to* 0 *in time* t.

Following Brammer [3 , p.347] and noting that the reachable sets $R(t) \triangleq \{\int_0^t e^{A(t-\tau)}Bu(\tau)d\tau \mid u(\tau) \in \Omega\}$, $t > 0$, are convex (even when Ω is not compact - cf. Section 8.2 above), it may be proved (we refrain from doing so here, but see [5]), that a necessary and sufficient condition (the condition is sufficient if $0 \in cl^{(1)}$(convexhull($B\Omega$))) for arbitrary-interval null-controllability of (8.1.1) is that there should exist for

1) cl(•) denotes the closure of the set (•).

each non-zero $v \in \mathbb{R}^n$ and for each time $t > 0$ a time τ, $0 \leqslant \tau \leqslant t$ and a $u \in \Omega$ such that $v'e^{A\tau}Bu > 0$. We shall make frequent use of this ob= servation. First we state the following proposition.

PROPOSITION 8.4.2 [6] *If Ω is bounded, then a necessary condition for arbitrary-interval null-controllability of system* (8.1.1) *is that* $0 \in \text{cl}(\text{convexhull }(B\Omega))$. □

Consequently in what follows we shall adopt the standing assumption

$$0 \in \text{cl}(\text{Convexhull}(B\Omega)).$$

THEOREM 8.4.3 *The system* $(8.1.1)/((A,B,\Omega))$ *is arbitrary interval null-controllable iff the system* $(A,B,\text{cl }(\text{conichull}(\text{convexhull}(\Omega))))$ *is arbitrary-interval null-controllable.*

PROOF In view of the above observation one has arbitrary-interval null-controllability (of the system (8.1.1)) iff $\forall\ v \in \mathbb{R}^n$, $v \neq 0$, and for all $t > 0$ there exist τ, $0 \leqslant \tau \leqslant t$, and $u \in \Omega$ such that $v'e^{A\tau}Bu > 0$. Now $v'e^{A\tau}Bu = u'B'e^{A'\tau}v$, so that one has arbitrary-interval null-controlla= bility of the system (8.1.1) iff for all $v \in \mathbb{R}^n$, $v \neq 0$, and for all $t > 0$ there exists τ, $0 \leqslant \tau \leqslant t$ such that $B'e^{A'\tau}v \notin \Omega^{*}$(2). Ω^* is always a closed and convex cone, so that due to a well known separation theorem we have that $B'e^{A'\tau}v \notin \Omega^*$ iff there exists $a \in \mathbb{R}^m$ such that $a'w^* \leqslant 0$ for all $w^* \in \Omega^*$ and $a'B'e^{A'\tau}v > 0$. Hence one has arbitrary-interval null-controllability of the system (8.1.1) iff one has arbitrary-interval null-controllability of the system (A,B,Ω^{**}). Now since $\Omega \subset \Omega^{**}$ we conclude that $\Omega \subset \text{cl}(\text{conichull}(\text{convexhull}(\Omega))) \subset \Omega^{**}$

Obviously controllability with the set $\text{cl}(\text{conichull}(\text{convechull}(\Omega)))$ as control restraint set implies controllability with Ω^{**}, which in turn implies controllability with Ω. □

---

2) We denote by $(\cdot)^*$ the polar of the set $(\cdot)$.

Hence when considering the arbitrary-interval null-controllability of system (8.1.1) one may use the ('nicer') control restraint set cl(conichull(convexhull(Ω))).

Also the following can be shown to hold.

PROPOSITION 8.4.4 [5] *A necessary condition for arbitrary-interval null-controllability of system* (8.1.1) *is that* cl(conichull(convexhull($B\Omega$))) (*or: the cone* cl(conichull(convexhull(Ω))) *if the matrix* B *is full rank*) *is not pointed.*

Thus for single-input systems, if $0 \notin \text{int}^{(3)}$(convexhull($\Omega$)) then the system is not arbitrary-interval null-controllable, as is illustrated by Example 8.4.1. Proposition 8.4.4 also motivates us to consider the largest subspace contained in the closed and convex cone $C \triangleq$ cl(conichull (convexhull($B\Omega$))). Indeed, let S be the largest subspace contained in C; an algorithm for computing S is given in [7]. In the next section we give a useful condition for arbitrary-interval null-controllability in terms of this subspace.

8.5 A CHARACTERIZATION OF ARBITRARY-INTERVAL NULL-CONTROLLABILITY IN TERMS OF THE SUBSPACE S

Notational Remark Given a set $\chi \in S^{\perp}$ ($\perp$ denotes the orthogonal complement), we shall denote by χ^0 the interior of the set χ relative to the subspace $S^{\perp} + \{0\}$. We shall denote by $\{A|S\}$ the smallest subspace invariant under A which contains the subspace S. It is well known that $\{A|S\} = S + AS + \ldots + A^{n-1}S$. We are now ready to state the following result.

---

3) int($\cdot$) denotes the interior of the set ($\cdot$).

THEOREM 8.5.1 [6] *A necessary condition for arbitrary-interval null-controllability of system* (8.1.1) *is that* $((B\,\mathrm{convexhull}(\Omega))^*)^0 \cap \{A|S\}^\perp = \{0\}$, *a sufficient condition for arbitrary-interval null-controllability is* $(B\,\mathrm{convexhull}(\Omega))^* \cap \{A|S\}^\perp = \{0\}$ *which is equivalent, if* Ω *is compact or* $0 \in B$ convexhull(Ω), *to the sufficient condition:* $0 \in \mathrm{int}(\mathrm{convexhull}(B\Omega) + \{A|S\})$. □

It is interesting to note that the assertion of Theorem 8.5.1 cannot in general be sharpened, as is shown by the following example.

EXAMPLE 8.5.2 Let

$$A = \begin{bmatrix} 0 & 0 & 0 & 1 \\ 0 & 1 & 0 & 0 \\ 0 & 1 & 0 & 1 \\ 0 & 0 & 0 & 0 \end{bmatrix}, \quad B = \begin{bmatrix} 1 & 0 & 0 & 0 \\ 0 & 1 & 0 & 0 \\ 0 & 0 & 1 & 0 \\ 0 & 0 & 0 & 1 \end{bmatrix}$$

and let the cone $\Omega = \text{linear span}\left(\begin{bmatrix} 0 \\ 0 \\ 0 \\ 1 \end{bmatrix}\right) + \{x | x \in \mathbb{R}^4, x_1 \geq 0, x_2^2 + x_3^2 \leq x_1^2, x_4 = 0\}$.

It turns out that the sufficient conditions of Theorem 8.4.1 do not hold, yet the system is arbitrary-interval null-controllable. Indeed, first note that the set $B\Omega$ is closed, so that

$$S = \text{linear span}\left(\begin{bmatrix} 0 \\ 0 \\ 0 \\ 1 \end{bmatrix}\right). \quad \text{Then}$$

$$\{A|S\} + B\Omega = \text{linear span}\left(\begin{bmatrix} 0 \\ 0 \\ 0 \\ 1 \end{bmatrix}, \begin{bmatrix} 1 \\ 0 \\ 1 \\ 0 \end{bmatrix}\right) + \{x | x \in \mathbb{R}^4, x_1 \geq 0, x_2^2 + x_3^2 \leq x_1^2, x_4 = 0\}.$$

It is readily verifiable that the vector $\begin{bmatrix} -1 \\ 0 \\ 1 \\ 0 \end{bmatrix} \notin \text{conichull}(\{A|S\} + B\Omega)$ so that $0 \notin \mathrm{int}(\{A|S\} + B\Omega)$. The above system is, however, arbitrary-interval null-controllable. Here,

$$v' e^{At} B u = v_1 u_1 + (e^t v_2 + e^t v_3 - v_3) u_2 + v_3 u_3 + (v_1 + v_3)\, t u_4.$$

Obviously, if $v_1 \neq -v_3$ then for all $t > 0$ there exists u_4 such that

$v'e^{At}Bu > 0$. Assume then that $v_1 + v_3 = 0$. Evidently, one should merely check for

$$v \in (B\Omega)^* = \{x \mid x \in \mathbb{R}^4,\ x_4 = 0, x_1 \leqslant 0,\ x_2^2 + x_3^2 \leqslant x_1^2\}.$$

The only vector in $(B\Omega)^*$ such that $v_1 + v_3 = 0$, is $v = \begin{bmatrix} -1 \\ 0 \\ 1 \\ 0 \end{bmatrix}$.

Then $v'e^{At}Bu = -u_1 + (e^t-1)u_2 + u_3$; $\begin{bmatrix} u_1 \\ u_2 \\ u_3 \\ u_4 \end{bmatrix} \in \Omega$.

Take $u_1 = 1$ and $u_2^2 + u_3^2 = 1$. Then $v'e^{At}Bu = -1 + (e^t-1)u_2 + \sqrt{1-u_2^2}$, $0 < u_2 \leqslant 1$. It is now readily verifiable that for all $t > 0$ there exists u_2, $0 \leqslant u_2 \leqslant 1$ such that $v'e^{At}Bu > 0$, so that in view of the observations in Section 8.4 the system is arbitrary-interval null-controllable.

In the following special case we can deduce a necessary and sufficient condition.

THEOREM 8.5.3 [6] *Assume that* $S^{\perp} \subset \mathbb{R}^2$. *Then a necessary and sufficient condition for arbitrary-interval null-controllability is* $(\text{convexhull}(B\Omega))^* \cap \{A|S\}^{\perp} = \{0\}$. *If in addition, the control restraint set* Ω *is compact or* $0 \in \text{convexhull}(\Omega)$, *then a necessary and sufficient condition for arbitrary-interval null-controllability is* □
$0 \in \text{int}(\text{convexhull}(B\Omega) + \{A|S\})$.

EXAMPLE 8.5.4 Let $A = \begin{bmatrix} 1 & 0 & 0 \\ -1 & -1 & 1 \\ -1 & 1 & -1 \end{bmatrix}$, $B = \begin{bmatrix} 1 & 0 & 0 \\ 0 & 1 & 0 \\ 0 & 0 & 1 \end{bmatrix}$,

and $\Omega = \{u \mid u \in \mathbb{R}^3,\ u_2 \geqslant 0,\ u_1 + u_2 \geqslant 0,\ u_1^2 + u_3^2 \leqslant 1,\ u_2^2 + u_3^2 \leqslant 1\}$

$C = \text{cl}(\text{conichull}(\text{convexhull}(B\Omega)) = \{x \mid x \in \mathbb{R}^3,\ x_2 \geqslant 0,\ x_1 + x_2 \geqslant 0\}$

$S = \text{linear span}(\begin{bmatrix} 0 \\ 0 \\ 1 \end{bmatrix})$, $\{A|S\} + C = \{x \mid x \in \mathbb{R}^3,\ x_2 \geqslant 0,\ x_1 + x_2 \geqslant 0\}$

$+ \text{ linear span } (\begin{bmatrix} 0 \\ 1 \\ 0 \end{bmatrix}), (\begin{bmatrix} 0 \\ 0 \\ 1 \end{bmatrix})$

Evidently, $0 \in \text{int}(\{A|S\} + C)$, so that the system is arbitrary-interval null-controllable.

8.6 APPLICATIONS

8.6.1 Continuity of the minimum-time function

In [8] the minimum-time optimal control problem is considered and the following assumptions are made:

(i) the control restraint set is a polyhedron;

(ii) the so-called *Pontryagin condition of general position* is assumed - in other words, if b is a vector co-linear with an edge of the polyhedron Ω, then $\{A|b\} = \mathbb{R}^n$;

(iii) $0 \in \Omega$, $0 \notin$ vertex (Ω).

It turns out [6] that the conditions (i)-(iii) imply that the system is arbitrary-interval null-controllable. In fact, via the notion of arbi= trary-interval null-controllability we obtain the following (minimal) condition for the continuity of the minimum-time function.

THEOREM 8.6.1 *Assume that the control restraint set* Ω *is compact. Then the minimum-time function* $T(x)$ *is continuous (in* x*) iff the system* (8.1.1) *is arbitrary-interval null-controllable.*

PROOF Sufficiency: Arbitrary-interval null-controllability implies that given $t > 0$ there exists an open neighbourhood of the origin, say $W(t)$, such that all points of $W(t)$ can be steered to 0 in time t. Owing to the continuous dependence of the solutions of differential equations on the initial values, we know that given a point x in the domain of defini= tion of $T(x)$, there exists a (sufficiently small) neighbourhood of x, say $\tilde{W}$, such that all points $y \in \tilde{W}$ can in time $T(x)$ be steered to $W(t)$, in other words: all points $y \in \tilde{W}$ can in time $T(x) + t$ be steered to the origin. Thus: given $t > 0, \exists$ a neighbourhood $\tilde{W}$ of x such that $T(y) - T(x) \leqslant t \; \forall \; y \in \tilde{W}$ (i.e. we have upper-semicontinuity of $T(x)$).

Also, lower-semicontinuity of $T(x)$ follows upon noting that the reach= able set (at time t, $t > 0$) is closed (since Ω is compact). Hence the minimum-time function $T(x)$ is continuous.

Necessity: Note that $T(x)$ is defined at 0, $T(0) = 0$. From the defini= tion of continuity we then have: for all $t > 0$, there exists an open neighbourhood of 0, say $W(t)$, such that all states in $W(t)$ are steered to 0 in time t, and this is precisely arbitrary-interval null-controlla= bility. □

8.6.2 Explicit conditions for controllability

In view of Theorem 8.4.3 we know (as far as arbitrary-interval null-controllability is concerned) that without loss of generality we can take the control restraint set $B\Omega$ to be a closed and convex cone, say C. Then in view of Theorem 8.5.1 we have the following simple sufficient condi= tion for cone-constrained (arbitrary-interval) controllability:

$$\{A|S\} + C = \mathbb{R}^n, \tag{8.6.1}$$

where S is the largest subspace contained in the control restraint cone C.

It turns out that the condition (8.6.1) allows one to explicitly compute a control function which will steer an initial state to the origin [9]. Moreover, if (8.6.1) holds it is also possible [10] to construct a piece= wise linear (linear-in-cones) feedback which stabilizes the system, as illustrated by the following example.

EXAMPLE 8.6.1

Let $A = \begin{bmatrix} 1 & 0 & 0 \\ -1 & -1 & 1 \\ -1 & 1 & -1 \end{bmatrix}$, $C = \{x | x \in \mathbb{R}^3,\ x_2 \geq 0, x_1 + x_2 \geq 0\}$.

In [10] it is shown that the following (nonlinear) feedback F(x) stabilizes the closed-loop system:

$$F(x) = \begin{bmatrix} -3x_1 \\ d(-3x_1) \\ -x_2 \end{bmatrix},$$

where $d : \mathbf{R}^1 \to \mathbf{R}^1$ is defined by $d(\sigma) = \begin{cases} 0 & \text{if } \sigma \geqslant 0 \\ -\sigma & \text{if } \sigma < 0. \end{cases}$

8.7 REFERENCES

[1] E.B. Lee, and L. Markus, *Foundations of Optimal Control Theory*. John Wiley, New York, 1967.

[2] S. Lefschetz, and J.P. La Salle, *Functional Analysis and Time Optimal Control*. Academic Press, New York, 1968.

[3] R.F. Brammer, Controllability in linear autonomous systems with positive controllers. *SIAM J. Cont.*, Vol 10, pp.339-353, 1972.

[4] O. Hãjek, A short proof of Brammer's theorem. Unpublished pre= print, 1975.

[5] M. Pachter and D.H. Jacobson, On the reachable sets in linear systems. *Applied Math. and Optimization*, Vol 5, pp 83-86, 1979.

[6] M. Pachter and D.H. Jacobson, Conditions for arbitrary-interval null-controllability and the continuity of the minimum-time function. NRIMS Technical Report WISK 210, CSIR, Pretoria, July 1976.

[7] V.Eckhardt, Theorems on the dimension of convex sets. *Linear Algebra and its Applications*, Vol 12, pp. 63-76, 1975.

[8] V.G. Boltyanskii, *Mathematical Methods of Optimal Control*. Holt, Rinehart, Winston Publishers, New York, 1971, pp 119-120.

[9] M. Pachter and D.H. Jacobson, Control with conic constraint set. *J. of Optimization Theory and Applications*, Vol 25, No 1, May 1978, pp 117-123.

[10] M. Pachter and D.H. Jacobson, Stabilization with conic control constraint set. *Int. J. of Control*, Vol 29, pp 125-132, 1979.

Additional Bibliography

Some of the material of Ref [5] is contained in the monograph:

D.H. Jacobson, *Extensions of Linear-Quadratic Control, Optimization and Matrix Theory*. Academic Press, London, England, 1977, ch.5.

Results on arbitrary-interval null-controllability which are different but nevertheless related to the work in [6], have been derived in

R.F. Brammer, Differential controllability and the solution of linear inequalities, Part I. IEEE Trans.Aut.Control, Vol AC-20, pp 128-131, 1975.

9. LINEAR-QUADRATIC PROBLEMS WITH CONICAL CONTROL SET

In general it is not possible, when there are controller constraints, to give explicit solutions for the optimal controllers of a linear control system with quadratic cost functional. In this lecture we determine explicit solutions for a restricted class of quadratic functionals defined on linear control systems with *conical* control constraints, and then, for the case of non-negative scalar control, develop necessary and sufficient conditions for the existence of optimal controls.

Interest in linear systems with conic control constraints was stimulated by the occurrence of such models in economic theory. Brammer [1] includes an analysis of an economic model with positive controllers, adapted from Paul A. Samuelson (Stiglitz [2]), while [3, Chap. 14] introduces LQ optimal control problems with positive controls in connection with economic stabilization policies.

9.1 A SPECIAL CLASS OF PROBLEMS WITH CONICAL CONTROL SET

We consider control systems of the form

$$\dot{x}(t) = A(t)x(t) + B(t)u(t) \qquad (9.1.1)$$

in which admissible controllers $u(\cdot)$ are defined and piecewise continuous on an interval $[t_0,T]$, and take values in a given closed cone $\Gamma \subseteq \mathbb{R}^m$:

$$u(t) \in \Gamma \text{ for all } t \in [t_0,T]. \qquad (9.1.2)$$

We take the final time T as given and fixed, whereas t_0 may vary. So far this is quite general, but we limit consideration to quadratic functionals of the form

$$J(t_0,x_0,u(\cdot)) = \int_{t_0}^{T} u'(t)R(t)u(t)dt + e(c'x(T))^2, \qquad (9.1.3)$$

where $x_0 = x(t_0)$, $e = \pm 1$, $c \in \mathbb{R}^n$ is non-zero, and where $R(\cdot)$ is con= tinuous and for all $t \leqslant T$, satisfies

$$u'R(t)u > 0 \text{ for all } u \in \Gamma \setminus \{0\}. \qquad (9.1.4)$$

The form (9.1.3) is thus special in two respects: the usual term $x'(t)Q(t)x(t)$ is missing in the integrand, and the final state matrix $Q_f = ecc'$ is of rank one, being positive or negative semidefinite according as $e = +1$ or -1. However, to relax these restrictions renders the approach presented below infeasible.

In the following sections we apply the fundamental sufficiency theorem of dynamic programming, i.e. Theorem 6.3.1 of Lecture 6, to completely solve the problem of minimizing (9.1.3) subject to (9.1.1) and (9.1.2). The results given here were originally derived in [4] following an entirely different approach.

Throughout, $\Phi(\cdot,\cdot)$ denotes the fundamental matrix solution associated with (9.1.1), so that

$$\frac{\partial\Phi(t,\tau)}{\partial t} = A(t)\Phi(t,\tau), \quad \frac{\partial\Phi(t,\tau)}{\partial\tau} = -\Phi(t,\tau)A(\tau), \quad \Phi(t,t) = I. \qquad (9.1.5)$$

9.2 SOLUTION OF THE H-J-B EQUATION

9.2.1 Some preliminary facts

In order to avoid interruption of the main argument we first prove some preliminary facts which we shall require.

For any $t \leqslant T$, $\xi \in \mathbb{R}^n$ and $u \in \mathbb{R}^m$ let

$$H(t,\xi,u) = 2\xi'B(t)u - u'R(t)u,$$

and consider the problem of maximizing $H(t,\xi,u)$ for fixed t and ξ, over $u \in \Gamma$. In consequence of the hypothesis (9.1.4) and the closedness of Γ this maximum is finite and is achieved - let

$$M(t,\xi) \triangleq \max_{u \in \Gamma} H(t,\xi,u).$$

LEMMA 9.2.1 *M is positively homogeneous of degree 2 in ξ. If $\hat{u} \in \Gamma$ is a maximizer for $H(t,\xi,\cdot)$, then $M(t,\xi) = \hat{u}'R(t)\hat{u}$, and for any $\lambda > 0$, $\lambda\hat{u}$ is a maximizer for $H(t,\lambda\xi,\cdot)$.*

PROOF For any $\lambda > 0$, we have $\lambda\Gamma = \Gamma$ (cone property), and hence, using the homogeneity of H in ξ,u jointly, we have

$$\begin{aligned} M(t,\lambda\xi) &= \max_{u \in \Gamma} H(t,\lambda\xi,u) = \max_{u \in \lambda\Gamma} H(t,\lambda\xi,u) \\ &= \max_{v \in \Gamma} H(t,\lambda\xi,\lambda v) = \lambda^2 \max_{v \in \Gamma} H(t,\xi,v) \\ &= \lambda^2 M(t,\xi). \end{aligned}$$

This proves the third claim as well as the homogeneity.

For the second claim, if $\hat{u}$ is a maximizer of $H(t,\xi,\cdot)$ in Γ, then, since $\beta\hat{u} \in \Gamma$ for all $\beta > 0$, we must have

$$\frac{d}{d\beta} H(t,\xi,\beta\hat{u}) = 0 \text{ at } \beta = 1.$$

This results in the equation

$$2\xi'B(t)\hat{u} - 2\hat{u}'R(t)\hat{u} = 0,$$

which can be written as $H(t,\xi,\hat{u}) - \hat{u}'R(t)\hat{u} = 0$, as required. □

LEMMA 9.2.2 *The function M is continuous.*

PROOF This is a consequence of (9.1.4) and the continuity of H. Sup=

pose $(t_n,\xi_n) \to (t_o,\xi_o)$; we show that then $M(t_n,\xi_n) \to M(t_o,\xi_o)$. If this were not so we could find $\varepsilon > 0$ and an infinite subsequence (t_n',ξ_n') for which

$$M(t_n',\xi_n') \notin (M(t_o,\xi_o) - \varepsilon,\ M(t_o,\xi_o) + \varepsilon). \qquad (9.2.1)$$

For each n, let $u_n' \in \Gamma$ be a maximizer of $H(t_n',\xi_n',\cdot)$ on Γ, and let $u_o \in \Gamma$ maximize $H(t_o,\xi_o,\cdot)$ on Γ. Then since

$$M(t_n',\xi_n') \geqslant H(t_n',\xi_n',u_o) \to H(t_o,\xi_o,u_o) = M(t_o,\xi_o),$$

we must, in view of (9.2.1), have

$$M(t_n',\xi_n') \geqslant M(t_o,\xi_o) + \varepsilon \qquad (9.2.2)$$

for all sufficiently large n.

Thus $M(t_n',\xi_n') = H(t_n',\xi_n',u_n')$ is bounded below, and it follows from (9.1.4) that the sequence $\{u_n'\} \subseteq \Gamma$ must be a bounded sequence, and hence has a convergent subsequence $u_n'' \to \bar{u}$ say. But then $(t_n'', \xi_n'',u_n'') \to (t_o,\xi_o,\bar{u})$, and by the continuity of H,

$$M(t_n'',\xi_n'') = H(t_n'',\xi_n'',u_n'') \to H(t_o,\xi_o,\bar{u}) \leqslant M(t_o,\xi_o).$$

This conclusion is clearly incompatible with (9.2.2), and the contra= diction proves the continuity of M. □

9.2.2 Solution of the H-J-B equation

In order to apply Theorem 6.3.1, we seek a class C^1 solution V(t,x) of H-J-B equation (6.3.1-2), which for the problem at hand becomes

$$-V_t(t,x) = \min_{u \in \Gamma} [u'Ru + V_x(t,x)(A(t)x + B(t)u)], \qquad (9.2.3)$$

with final condition

$$V(T,x) = e(c'x)^2. \qquad (9.2.4)$$

Should we succeed in finding our solution, then according to Theorem 6.3.1 the minimum value of $J(t_0,x_0,u(\cdot))$ over all admissible controllers ((t_0,x_0) being held fixed) would be $V(t_0,x_0)$. Using the 'variation of constants' formula (1.2.4) for the solution of (9.1.1) with initial condition $x(t_0) = x_0$, J may be written as

$$J(t_0,x_0,u(\cdot)) = \int_{t_0}^{T} u'(t)R(t)u(t)dt + e[c'\Phi(T,t_0)x_0 + \int_{t_0}^{T} c'\Phi(T,\tau)B(\tau)u(\tau)d\tau]^2;$$

showing that J depends on x_0 only through the linear form $c'\Phi(T,t_0)x_0$ - i.e. all initial states x_0 which give the same value to this linear form also give the same value to $J(t_0,x_0,u(\cdot))$ (for given t_0, $u(\cdot)$). It follows that the solution of (9.2.3-4) which we seek may be assumed to be of the form

$$V(t,x) = \psi(t,\alpha) \text{ with } \alpha = c'\Phi(T,t)x. \qquad (9.2.5)$$

The H-J-B equation then becomes

$$-\psi_t + \psi_\alpha\, c'\Phi(T,t)A(t)x$$

$$= \min_{u \in \Gamma}[u'Ru + \psi_\alpha c'\Phi(T,t)(A(t)x + B(t)u)],$$

whence

$$-\psi_t = \min_{u \in \Gamma} [u'Ru + \psi_\alpha\, c'\Phi(T,t)B(t)u].$$

$$= - \max_{u \in \Gamma}[- \psi_\alpha c'\Phi(T,t)B(t)u - u'Ru],$$

or finally

$$\psi_t = M(t,-\tfrac{1}{2}\psi_\alpha\Phi'(T,t)c). \qquad (9.2.6)$$

Prompted by the final-time condition (9.2.4), and by the homogeneity of M, we try a solution of the form

$$\psi = e\, p^2(t)\alpha^2 \quad \text{with } p(T) = 1. \tag{9.2.7}$$

Substituted in (9.2.6) this leads to

$$2p(t)\dot{p}(t)\alpha^2 = M(t, -\, ep^2(t)\alpha\Phi'(T,t)c) \tag{9.2.8}$$

$$= \begin{cases} (ep^2(t)\alpha)^2 M(t,-\Phi'(T,t)c) & \text{if } e\,\alpha \geqslant 0 \\ (ep^2(t)\alpha)^2 M(t,\Phi'(T,t)c) & \text{if } e\,\alpha \leqslant 0, \end{cases}$$

which reduces to

$$\dot{p}(t) = \begin{cases} \tfrac{1}{2} ep^3(t)M(t,-\Phi'(T,t)c) & \text{if } e\,\alpha \geqslant 0 \\ \tfrac{1}{2} ep^3(t)M(t,\Phi'(T,t)c) & \text{if } e\,\alpha \leqslant 0. \end{cases}$$

Noting the final-time condition in (9.2.7), these equations integrate to give

$$p(t) = \begin{cases} p_-(t) \triangleq [1 + eI_-(t)]^{-\frac{1}{2}} & \text{if } e\,\alpha \geqslant 0 \\ p_+(t) \triangleq [1 + eI_+(t)]^{-\frac{1}{2}} & \text{if } e\,\alpha \leqslant 0, \end{cases}$$

where

$$I_+(t) = \int_t^T M(\tau,\Phi'(T,\tau)c)d\tau, \qquad I_-(t) = \int_t^T M(\tau,-\Phi'(T,\tau)c)d\tau.$$

Our solution $V(t,x) = \psi(t,c'\Phi(T,t)x)$ is thus

$$V(t,x) = \begin{cases} e[1 + eI_-(t)]^{-1}(c'\Phi(T,t)x)^2 & \text{if } ec'\Phi(T,t)x \geqslant 0 \\ e[1 + eI_+(t)]^{-1}(c'\Phi(T,t)x)^2 & \text{if } ec'\Phi(T,t)x \leqslant 0. \end{cases} \tag{9.2.9}$$

It is easily seen that this solution is of class C^1 and satisfies the final-time condition (9.2.4). The optimal controls are those which afford the maximum value in (9.2.8), and so, using Lemma 9.2.1 again, are given by

$$u = \begin{cases} ep_-^2(t)(c'\Phi(T,t)x)u_-(t) & \text{if } ec'\Phi(T,t)x \geqslant 0 \\ \\ ep_+^2(t)(c'\Phi(T,t)x)u_+(t) & \text{if } ec'\Phi(T,t)x \leqslant 0, \end{cases} \qquad (9.2.10)$$

where $u_-(t)$ and $u_+(t)$ respectively maximize the specific functions

$$H(t,-\Phi'(T,t)c,\cdot) \text{ and } H(t,\Phi'(T,t)c,\cdot) \qquad (9.2.11)$$

over the control cone Γ. Note that for each time t, these controls are piecewise linear, being in fact linear in each of the half-spaces separated by the hyperplane

$$c'\Phi(T,t)x = 0.$$

9.2.3 Focal times

Since M is non-negative, the integrals $I_+(t)$, $I_-(t)$ are non-negative and non-decreasing as t decreases from t = T. Hence when e = +1, the solution (9.2.9-10) is defined for all $t \leqslant T$. However, when e = -1 (so that J is not *a priori* non-negative) the solution (9.2.9) 'blows up' as either $I_+(t)$ or $I_-(t)$ tends to 1, and is thus defined only for $t > t^*$, where

$$t^* = \inf\{t_0 < T \mid I_+(t_0) < 1 \text{ and } I_-(t_0) < 1\}.$$

This instant plays the role of a focal time, for it can be shown (see [4]) that for any $t_0 < t^*$, there exist initial states x_0 for which $J(t_0,x_0,\cdot)$ is unbounded below on the class of admissible controllers .

On the other hand, if we impose the homogeneous initial condition $x_0 = 0$, it can be shown (see [4]) that $J(t_0, 0, \cdot)$ is non-negative on the class of admissible controllers for all $t_0 > t^{**} = \inf\{t_0 < T \mid I_+(t) \leq 1$ and $I_-(t) \leq 1\}$.

9.2.4 An alternative computational scheme

As presented above, the solution of the problem requires the determination of the two special controllers $u_+(\cdot)$, $u_-(\cdot)$ by maximization of the functions (9.2.11) over the control cone Γ - this requires the prior determination of the vector function $\Phi'(T,t)c$ - and then the evaluation by quadrature of the integrals $I_+(\cdot)$, $I_-(\cdot)$. However a more streamlined procedure can be derived from (9.2.8) as follows. We use the homogeneity of M in a different way to derive from (9.2.8) the equation

$$2p(t)\dot{p}(t)\alpha^2 = \begin{cases} (ep(t)\alpha)^2 M(t,-p(t)\Phi'(T,t)c) & \text{if } ep(t)\alpha \geq 0 \\ \\ (ep(t)\alpha)^2 M(t,p(t)\Phi'(T,t)c) & \text{if } ep(t)\alpha \leq 0, \end{cases}$$

which reduces to

$$\dot{p}(t) = \begin{cases} \tfrac{1}{2}ep(t)M(t,-p(t)\Phi'(T,t)c) & \text{if } ep(t)c'\Phi(T,t)x \geq 0 \\ \\ \tfrac{1}{2}ep(t)M(t,p(t)\Phi'(T,t)c) & \text{if } ep(t)c'\Phi(T,t)x \leq 0. \end{cases} \qquad (9.2.12)$$

This suggests the introduction of the vector function

$$\xi(t) = p(t)\Phi'(T,t)c \qquad (9.2.13)$$

for which, using (9.2.12), we find

$$\dot{\xi}(t) = \begin{cases} -A'(t)\xi(t) + \tfrac{1}{2}eM(t,-\xi(t))\xi(t) & \text{if } e\xi'(t)x \geqslant 0 \\ \\ -A'(t)\xi(t) + \tfrac{1}{2}eM(t,\xi(t))\xi(t) & \text{if } e\xi'(t)x \leqslant 0. \end{cases} \qquad (9.2.14)$$

It follows that our solution may be written as

$$V(t,x) = \begin{cases} e(\xi_-'(t)x)^2 & \text{if } e\xi_-'(t)x \geqslant 0 \\ \\ e(\xi_+'(t)x)^2 & \text{if } e\xi_+'(t)x \leqslant 0, \end{cases}$$

where $\xi_-(\cdot)$ satisfies the first option in (9.2.14), and $\xi_+(\cdot)$ satisfies the second, commencing with the final condition

$$\xi_+(T) = \xi_-(T) = c.$$

Furthermore it is easily checked that the optimal feedback is

$$u = \begin{cases} e(\xi_-'(t)x)v_-(t) & \text{if } e\xi_-'(t)x \geqslant 0 \\ \\ -e(\xi_+'(t)x)v_+(t) & \text{if } e\xi_+'(t)x \leqslant 0, \end{cases}$$

where $v_-(t)$, $v_+(t)$ respectively maximize

$$H(t,-\xi_-(t),\cdot) \text{ and } H(t,\xi_+(t),\cdot) \qquad (9.2.15)$$

over Γ. Thus knowledge of $\xi_-(t)$(or $\xi_+(t)$) at time t permits evaluation by these maximizations of the optimal control law at time t, while the maximum values $M(t, -\xi_-(t))$ (or $M(t,\xi_+(t))$) are precisely what is re= quired in (9.2.14) to evaluate $\dot{\xi}_-(t)$ (or $\dot{\xi}_+(t)$).

This ends the first part of this Lecture; the second part, in a quite different way, tackles the problem of one-sided controller constraints.

9.3 THE CASE OF NON-NEGATIVE SCALAR CONTROL

In the remaining two sections we consider the autonomous linear control system

$$\dot{x} = Ax + bu, \quad 0 \leqslant t \leqslant T$$

with a *scalar* control $u \in \mathbb{R}$ constrained to allow only non-negative values:

$$u \geqslant 0.$$

We consider the problem of minimizing cost criteria of the form

$$J(x_0,u(\cdot)) = \int_0^T [u^2 + x'(t)Qx(t)]\,dt + x'(T)Q_f x(T),$$

in which neither Q nor Q_f is assumed to be positive semidefinite, and we derive necessary and sufficient conditions for the existence, for all initial states $x_0 \in \mathbb{R}^n$, of controls which minimize $J(x_0,\cdot)$.

Basically, our approach is to strengthen Pontryagin's necessary condi= tions (for an optimum for all $x_0 \in \mathbb{R}^n$) into sufficiency conditions - a method also referred to as 'regular synthesis' - see in particular [5, ch. IV] and also [6, sect. 12], [7]. In addition, we apply methods of nonlinear analysis in $\mathbb{R}^n$, and the reader is urged to consult Appendix A in [8].

The results we obtain (on the existence of solutions, conjugate/focal times) are explicit. We illustrate the theorems by way of examples, and in addition we show that our necessary/sufficient conditions be= come the classical necessary and sufficient conditions in the special case of the unconstrained control problems; we refer the reader to [8] for the proofs of the theorems and generalizations.

We first introduce the following function $f:\mathbb{R}^1 \to \mathbb{R}^1$. Let $f: \mathbb{R}^1 \to \mathbb{R}^1$ be given by

$$f(v) \triangleq \begin{cases} v \text{ if } v \geqslant 0 \\ \\ 0 \text{ if } v < 0 \end{cases}, \quad v \in \mathbb{R}^1,$$

and consider the mappings $F_\tau: \mathbb{R}^n \to \mathbb{R}^n$, $0 < \tau \leqslant T$, defined as follows: given $z \in \mathbb{R}^n$, solve the 'canonical' system

$$\dot{x} = Ax + bf(-b'\lambda) \qquad 0 \leqslant t \leqslant \tau$$
$$\dot{\lambda} = - Qx - A'\lambda$$

with the 'initial' condition $x(\tau) = z$, $\lambda(\tau) = Q_f z$, and set $F_\tau(z) = x(0)$.

Note that the mapping F_τ so defined is positively homogeneous and con= tinuous.

Employing the 'regular synthesis' argument and an 'inverse function' type of argument in $\mathbb{R}^n$, the following can be shown to hold.

THEOREM 9.3.1 *A necessary condition for the existence of a solution is that the mappings* F_τ *are onto for all* $0 < \tau \leqslant T$. *A sufficient condition for the existence of a solution is that the mappings* F_τ *are onto and one-to-one for all* $0 < \tau \leqslant T$. □

We are thus faced with the problem of verifying whether the positively homogeneous and continuous mappings $F_\tau: \mathbb{R}^n \to \mathbb{R}^n$, $0 < \tau \leqslant T$, are surjective and /or bijective. If we use the notation

$$T_1 \triangleq \sup_{T > 0} (\{T | F_\tau \text{ is a surjection } \forall\ 0 < \tau \leqslant T\}),$$

$$T_2 \triangleq \sup_{T > 0} (\{T | F_\tau \text{ is a bijection } \forall\ 0 < \tau \leqslant T\})$$

(obviously $T_2 \leqslant T_1$) we know that if $T < T_2$, there exists a unique optimal (feedback) solution for all initial states $x_0 \in \mathbb{R}^n$; and if $T > T_1$, there does not exist an optimal solution for some $x_0 \in \mathbb{R}^n$.

We shall next require the following definitions:

$$S^{n-1} \triangleq \{x | x \in \mathbb{R}^n, |x| = 1\},$$

$$\alpha(\tau) \triangleq \min_{x \in S^{n-1}} |F_\tau(x)|,$$

$$\alpha_T \triangleq \inf_{0 < \tau \leqslant T} \alpha(\tau).$$

By employing ideas from the theory of the 'degree of a mapping' we were able to obtain conditions for surjectiveness/injectiveness of a mapping, as follows.

THEOREM 9.3.2 *Let* $\tilde{T}$ *be the smallest positive real number such that* $\alpha_{\tilde{T}} = 0$. *If* $T < \tilde{T}$, *then the mappings* F_τ, $0 \leqslant \tau \leqslant T$ *are surjective (i.e. we know that Pontryagin's necessary condition holds). If, however,* $T \geqslant \tilde{T}$, *there exists a* τ, $0 < \tau \leqslant T$, *such that* F_τ *is not injective.* □

Alternatively, define

$$\gamma(\tau) \triangleq \min_{x \in S^{n-1}} x'F_\tau(x)$$

and

$$\gamma_T \triangleq \inf_{0 < \tau \leqslant T} \gamma(\tau).$$

By invoking a coerciveness-type of theorem in $\mathbb{R}^n$, it can be shown that the following theorem holds.

THEOREM 9.3.3 *Let* $\tilde{\tilde{T}}$ *be the smallest positive real number such that* $\gamma_{\tilde{\tilde{T}}} = 0$. *If* $T < \tilde{\tilde{T}}$, *then the mappings* F_τ, $0 < \tau \leqslant T$ *are surjective.* □

We now apply these results to the classical, control-unconstrained problem. In this case the mapping F_τ, $0 < \tau \leqslant T$ is linear in z and is given by

$$F_\tau(z) = (I_n,0)e^{-\begin{pmatrix} A & -bb' \\ -Q & -A' \end{pmatrix}\tau} \begin{pmatrix} z \\ Q_f z \end{pmatrix}$$

i.e. there exists an n×n matrix $M(\tau)$ such that

$$F_\tau(z) = M(\tau)z$$

Since any linear map that maps $\mathbb{R}^n$ into $\mathbb{R}^n$ is surjective iff it is bijective, there is now no gap in the set of necessary/sufficient conditions of Theorem 9.3.1, which therefore yields the following corollary.

COROLLARY 9.3.4 *A necessary and sufficient condition for the existence of a solution in the control-unconstrained case, for all initial states* $x_0 \in \mathbb{R}^n$, *is that*

$$\det(M(\tau)) \neq 0 \text{ for all } 0 < \tau \leqslant T. \qquad \square$$

Thus, denote by $\bar{T}$ the smallest positive real number such that $\det(M(\bar{T}))=0$. Then $T_1 = T_2 = \bar{T}$.

In addition, in view of Lemma 1 in [9], and since any linear map mapping $\mathbb{R}^n$ into $\mathbb{R}^n$ is injective iff the homogeneous system has the unique solution 0, it is obvious that the following also holds:

$$\bar{T} = \sup_{T>0} (\{T \mid \text{there exists a unique solution } u^* = 0 \text{ on } [0,T] \text{ to the optimal control problem with } x_0 = 0\}).$$

Hence $\bar{T}$ is in fact the conjugate/focal time. We have thus recovered the classical disconjugacy results for the special case of the control-unconstrained linear-quadratic optimal control problem (see, for example [10 , ch. 5] or [9]).

9.4 THE SPECIAL CASE Q = 0

If in the quadratic performance criterion Q = 0, the adjoint equations in the canonical system are independent of the state variables, it is then possible to obtain explicit conditions for the existence of solu= tions.

Specifically, we have

$$\lambda(t) = e^{A'(T-t)}\lambda(T), \; 0 \leqslant t \leqslant T$$

and

$$\lambda(T) = Q_f z.$$

Hence

$$\lambda(t) = e^{A'(T-t)}Q_f z, \quad 0 \leqslant t \leqslant T$$

so that

$$u^*(t) = f(-b'\lambda(t)) = f(-b'e^{A'(T-t)}Q_f z), \quad 0 \leqslant t \leqslant T.$$

Then by the variation-of-parameters formula,

$$x(\tau) = e^{A\tau}x_0 + \int_0^\tau e^{A\sigma}bf(-b'e^{A'\sigma}Q_f z)d\sigma$$

and hence the mapping F_τ, $0 < \tau \leqslant T$ is given explicitly by

$$z \longmapsto e^{-A\tau}[z - \int_0^\tau e^{A\sigma}bf(-b'e^{A'\sigma}Q_f z)d\sigma]. \qquad (9.4.1)$$

Using the explicit form of the mapping F_τ given here, and a 'monotone mapping'-type of result, we are able to show the following.

THEOREM 9.4.1 Let

$$F(x,\bar{x};t) \triangleq \int_0^t [(x-\bar{x})'e^{A\sigma}bf(-b'e^{A'\sigma}Q_f x)-f(-b'e^{A'\sigma}Q_f\bar{x})]d\sigma.$$

A sufficient condition for the existence of a solution to our optimal control problem, for all $x_0 \in \mathbb{R}^n$, *is that*

(a) $$g_1(t) \triangleq \sup_{|x| \leqslant 1, |\bar{x}| \leqslant 1,\ x \neq \bar{x}} (F(x,\bar{x};t)/|x-\bar{x}|^2) < 1$$

for all $0 \leqslant t \leqslant T$;

or

(b) $$g_2(t) \triangleq \max_{|x| \leqslant 1, |\bar{x}| \leqslant 1} (F(x,\bar{x};t)) \leqslant 0 \text{ for all } 0 \leqslant t \leqslant T.$$ □

The condition (b) is simpler than condition (a). Theorem 9.4.1(a), however, yields the following corollary.

COROLLARY 9.4.2 If

$$\int_0^T |e^{A\sigma}B|^2 d\sigma < 1/|Q_f|$$

then there exists a solution to our optimal control problem for all initial states $x_0 \in \mathbb{R}^n$. □

Therefore we know that if $|Q_f|$ is sufficiently small, existence of a solution is guaranteed.

In the classical, control-unconstrained situation, parts (a) and (b) of Theorem 9.4.1 yield the following corollary.

COROLLARY 9.4.3 Let

$$W(t) \triangleq \int_0^t e^{-A\tau}BB'e^{-A'\tau}d\tau.$$

Then a sufficient condition for the existence of a solution to our optimal control problem, for all $x_0 \in \mathbb{R}^n$, *is the following.*

(a) *for all* $0 \leqslant t \leqslant T$

$$z'(I + e^{At}W(t)e^{A't}Q_f)z > 0 \text{ for all } z \in \mathbb{R}^n,\ z \neq 0. \tag{9.4.2}$$

If, in addition, the pair (A,B) *is controllable, then condition* (9.4.2) *is equivalent to*

$$Q_f + e^{-A't}W^{-1}(t)e^{-At} > 0 \text{ for all } 0 < t \leqslant T; \tag{9.4.3}$$

or,

(b) *for all* $0 \leqslant t \leqslant T$

$$z'e^{At}W(t)e^{A't}Q_f z \geqslant 0 \text{ for all } z \in \mathbb{R}^n.$$

PROOFS

(a) Theorem 9.4.1(a) gives, for all $0 \leqslant t \leqslant T$,

$$1 > g_1(t) = -[z' \int_0^t e^{A\sigma}BB'e^{A'\sigma}d\sigma Q_f z]/z'z \text{ for all } z \in \mathbb{R}^n \quad z \neq 0.$$

Since

$$\int_0^t e^{A\sigma}BB'e^{A'\sigma}d\sigma = e^{At}W(t)e^{A't}$$

we immediately obtain (9.4.2). If in addition the pair (A,B) is controllable, then the matrix W(t) is positive definite. Since then $I + e^{At}W(t)e^{A't}Q_f$ is non-singular iff $Q_f + e^{-A't}W^{-1}(t)e^{-At}$ is non-singular, and since $W^{-1}(t)$ is positive definite and very large for small t, W(0) = 0 and z.z > 0 if $z \neq 0$, we conclude that (9.4.2) holds for all $0 \leqslant t \leqslant T$ iff (9.4.3) holds for all $0 < t \leqslant T$.

(b) This follows directly from Theorem 9.4.1(b). □

Thus it should be noted that with unconstrained control, if the pair (A,B) is controllable, then the sufficiency condition (9.4.3) in Corollary 9.4.3(a) is the well-known necessary and sufficient condition of Riccati type ([11], Theorem 2, p 134) for the existence of a solution to our optimal control problem for all initial states $x_o \in \mathbb{R}^n$.

Theorems 9.3.2 and 9.3.3 are also readily applicable, the mappings F_τ being given explicitly by equation (9.4.1). Specifically, the positive number $\tilde{T}$ in Theorem 9.3.2 is given explicitly by

$$\tilde{T} = \inf_{T>0}(\{T | \max_{|x|=1} |\int_o^T e^{A\sigma} bf(-b'e^{A'\sigma}Q_f x)d\sigma| = 1\}).$$

Similarly, the positive number $\tilde{\tilde{T}}$ in Theorem 9.3.3 is now given explicitly by

$$\tilde{\tilde{T}} = \inf_{T>0} (\{T | \max_{|x|=1} \int_o^T x'e^{A\sigma}bf(-b'e^{A'\sigma}Q_f x)d\sigma = 1\}).$$

In the classical problem with unconstrained control, Theorems 9.3.2 and 9.3.3 thus yield the following sufficient conditions for the existence of a solution for all $x_o \in \mathbb{R}^n$:

$$\min_{(x,t) \in S^{n-1}\times[0,T]} |(I + e^{At}W(t)e^{A't}Q_f)x| > 0,$$

which is equivalent to the condition of Corollary 9.4.3(a), which in turn is equivalent to the classical necessary and sufficient condition of Theorem 2., p.134 of [11]; and also to

$$\min_{|x|=1} x'e^{At}W(t)e^{A't}Q_f x > -1 \text{ for all } 0 \leqslant t \leqslant T.$$

It turns out that the complexity of the problem is significantly reduced if the matrix Q_f is of low rank. Indeed, in view of the special structure of our mappings (9.4.1) we have the following proposition.

PROPOSITION 9.4.4 *Let* rank H=r *and let* $K \in \mathbb{R}^{n\times r}$ *such that* $\text{Range}(K) = (\text{Ker}(Q_f))^\perp$ *and* $K'K = I_r$ *(we here denote by* Ker(·) *the null space of a matrix, and by* I_r *the identity matrix in* $\mathbb{R}^r$*). Thus the mapping*

(9.4.1) *is a bijection iff the mapping* $\tilde{F}_T\colon \mathbb{R}^r \to \mathbb{R}^r$ *is a bijection, where*

$$\tilde{F}_T(\eta) \triangleq \eta - K' \int_0^T e^{A\sigma} b f(-b' e^{A'\sigma} Q_f K \eta)\,d\sigma, \quad \eta \in \mathbb{R}^r. \qquad (9.4.4.) \qquad \square$$

Hence, if $r = 1$ (i.e. if $Q_f = qcc'$, where $q \in \mathbb{R}^1$ and $c \in \mathbb{R}^n$; i.e. our system has a single output), then the necessary and sufficient con= dition for the mapping $\tilde{F}_T$ to be a bijection is simply

$$\tilde{F}_T(1)\cdot\tilde{F}_T(-1) < 0. \qquad (9.4.5)$$

If $r = 2$ (i.e. if we have a dual-output system), then the mapping $\tilde{F}_T$ is onto iff for all $\theta \in (0,2\pi]$, there exists $r > 0$ such that $(r\cos\theta, r\sin\theta) \in \tilde{F}_T(S^1)$. Hence Pontryagin's principle now becomes: for all $\theta \in (0,2\pi]$, there exists $r > 0$ such that $(r\cos\theta, r\sin\theta) \in \tilde{F}_T(S^1)$. In addition, the conditions for the mapping $\tilde{F}_T$ to be a bijection are now:

(i) $F_T(S^1)$ is a Jordan curve;

(ii) $0 \in$ int(bounded part of the plane determined by the Jordan curve $F_T(S^1)$);

(iii) if $x,y, \in \tilde{F}_T(S^1)$, $x \neq y$, then $\nexists\, \alpha \geqslant 0$ such that $x = \alpha y$.

EXAMPLE 9.4.5 Consider the harmonic oscillator

$\ddot{x} + x = u$, $x(0) = x_0$, $0 \leqslant t \leqslant T$; $u \geqslant 0$ and

$$J(u,x_0) = \tfrac{1}{2}[-qx^2(T) + \int_0^T u^2(t)dt].$$

Setting

$$x = \begin{bmatrix} x_1 \\ x_2 \end{bmatrix}, \text{ we have } A = \begin{bmatrix} 0 & 1 \\ -1 & 0 \end{bmatrix}, \; b = \begin{bmatrix} 0 \\ 1 \end{bmatrix}, \quad Q_f = \begin{bmatrix} -q & 0 \\ 0 & 0 \end{bmatrix}.$$

Hence rank $(Q_f) = 1$ and $K = \binom{1}{0}$ so that (see (9.4.4)):

$$\tilde{F}_T(\eta) = \eta - (1,0)\int_0^T \begin{pmatrix}\cos\sigma & \sin\sigma\\ -\sin\sigma & \cos\sigma\end{pmatrix}\begin{pmatrix}0\\1\end{pmatrix} f\left(-(0,1)\begin{pmatrix}\cos\sigma & -\sin\sigma\\ \sin\sigma & \cos\sigma\end{pmatrix}\begin{pmatrix}-q & 0\\ 0 & 0\end{pmatrix}\begin{pmatrix}1\\0\end{pmatrix}\eta\right)d\sigma$$

$$= \eta - \int_0^T \sin\sigma\, f(q\eta\sin\sigma)d\sigma.$$

Assume that $q = 2/\pi$. Then (see (9.4.5))

$$\tilde{F}_T(1)\tilde{F}_T(-1) = [1 - \frac{2}{\pi}\int_0^T \sin\sigma\, f(\sin\sigma)d\sigma]\,[-1 - \frac{2}{\pi}\int_0^T \sin\sigma f(-\sin\sigma)d\sigma],$$

and since $\int_0^\pi \sin^2\sigma\, d\sigma = \pi/2$, we conclude that

$$T_1 = T_2 = \pi.$$

EXAMPLE 9.4.6 Again, $A = \begin{pmatrix}0 & 1\\ -1 & 0\end{pmatrix}$, $b = \begin{pmatrix}0\\1\end{pmatrix}$; however $Q_f = \begin{pmatrix}-q & 0\\ 0 & -q\end{pmatrix}$ (i.e. rank $(Q_f) = 2$).

Upon parametrizing $x = \begin{pmatrix}\cos\alpha\\ \sin\alpha\end{pmatrix}$, $0 < \alpha \leqslant 2\pi$, we have that our mapping is now

$$\begin{pmatrix}\cos\alpha\\ \sin\alpha\end{pmatrix} \longmapsto \begin{pmatrix}\cos\alpha\\ \sin\alpha\end{pmatrix} - q\int_0^T \begin{pmatrix}\sin\sigma\\ \cos\sigma\end{pmatrix} f(\sin(\alpha+\sigma))d\sigma.$$

Set $\alpha+\sigma = t$. Then

$$\begin{pmatrix}\cos\alpha\\ \sin\alpha\end{pmatrix} \to \begin{pmatrix}\cos\alpha\\ \sin\alpha\end{pmatrix} - q\int_\alpha^{T+\alpha} \begin{pmatrix}\sin(t-\alpha)\\ \cos(t-\alpha)\end{pmatrix} f(\sin t)dt.$$

Now

$$\int \sin(t-\alpha)\sin t\, dt = \frac{t\cos\alpha}{2} - \tfrac{1}{4}\sin(2t-\alpha) \text{ and}$$

$$\int \cos(t-\alpha)\sin t\, dt = \frac{t\sin\alpha}{2} - \tfrac{1}{4}\cos(2t-\alpha);$$

so that if for example $T = 2\pi$, since

$$\int_{\alpha}^{2\pi+\alpha}\begin{bmatrix}\sin(t-\alpha)\\ \cos(t-\alpha)\end{bmatrix} f(\sin t)\,dt = \begin{cases} \int_{\alpha}^{\pi}\sin(t-\alpha)\sin t\,dt + \int_{2\pi}^{2\pi+\alpha}\sin(t-\alpha)\sin t\,dt = \frac{\pi}{2}\cos\alpha \\ \qquad\qquad \text{if } 0 \leqslant \alpha < \pi \\ \int_{\alpha}^{\pi}\cos(t-\alpha)\sin t\,dt + \int_{2\pi}^{2\pi+\alpha}\cos(t-\alpha)\sin t\,dt = \frac{\pi}{2}\sin\alpha \\ \int_{2\pi}^{3\pi}\sin(t-\alpha)\sin t\,dt = \frac{\pi}{2}\cos\alpha \\ \qquad\qquad \text{if } \pi \leqslant \alpha < 2\pi \\ \int_{2\pi}^{3\pi}\cos(t-\alpha)\sin t\,dt = \frac{\pi}{2}\sin\alpha \end{cases}$$

we have

$$\begin{bmatrix}\cos\alpha\\ \sin\alpha\end{bmatrix} \to \left[1 - q\,\frac{\pi}{2}\right]\begin{bmatrix}\cos\alpha\\ \sin\alpha\end{bmatrix} \Rightarrow \quad q \not\geqslant \frac{2}{\pi}$$

(indeed, if $q \geqslant \frac{2}{\pi}$ this would imply that q can also equal $\frac{2}{\pi}$, which however is impossible). Similarly, if $T = 4\pi \Rightarrow \; q \not\geqslant \frac{1}{\pi}$.

If, for example, q = 1,5/π, then

$F_{3.0}\big|_{S^1}(S^1)$ is given by

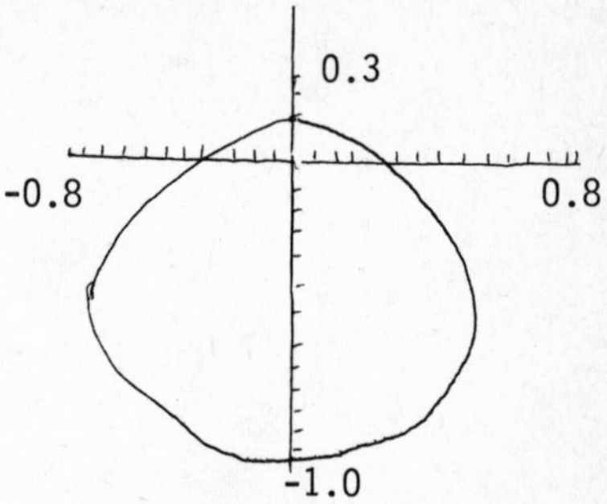

$F_{6.8}\big|_{S^1}(S^1)$ is given by

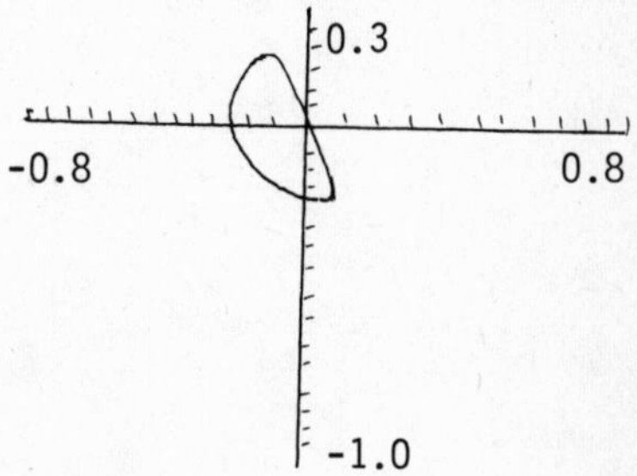

$F_{6.9}\big|_{S^1}(S^1)$ is given by

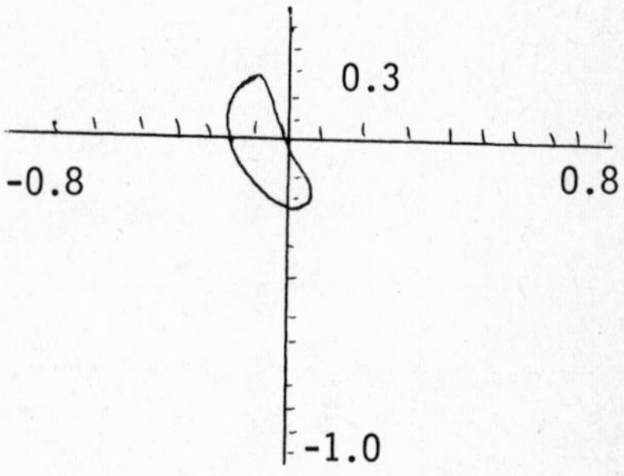

and

$F_{8.0}\big|_{S^1}(S^1)$ is given by

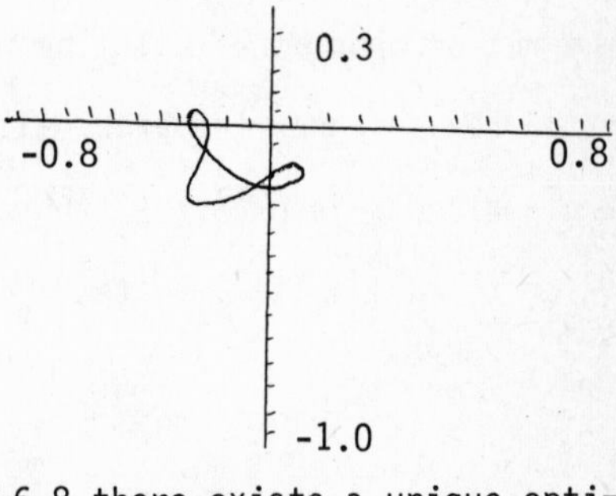

Hence we know that for q = 1.5/π and $T \leqslant 6.8$ there exists a unique opti= mal solution for all initial states $x_o \in R^n$.

In fact, we have

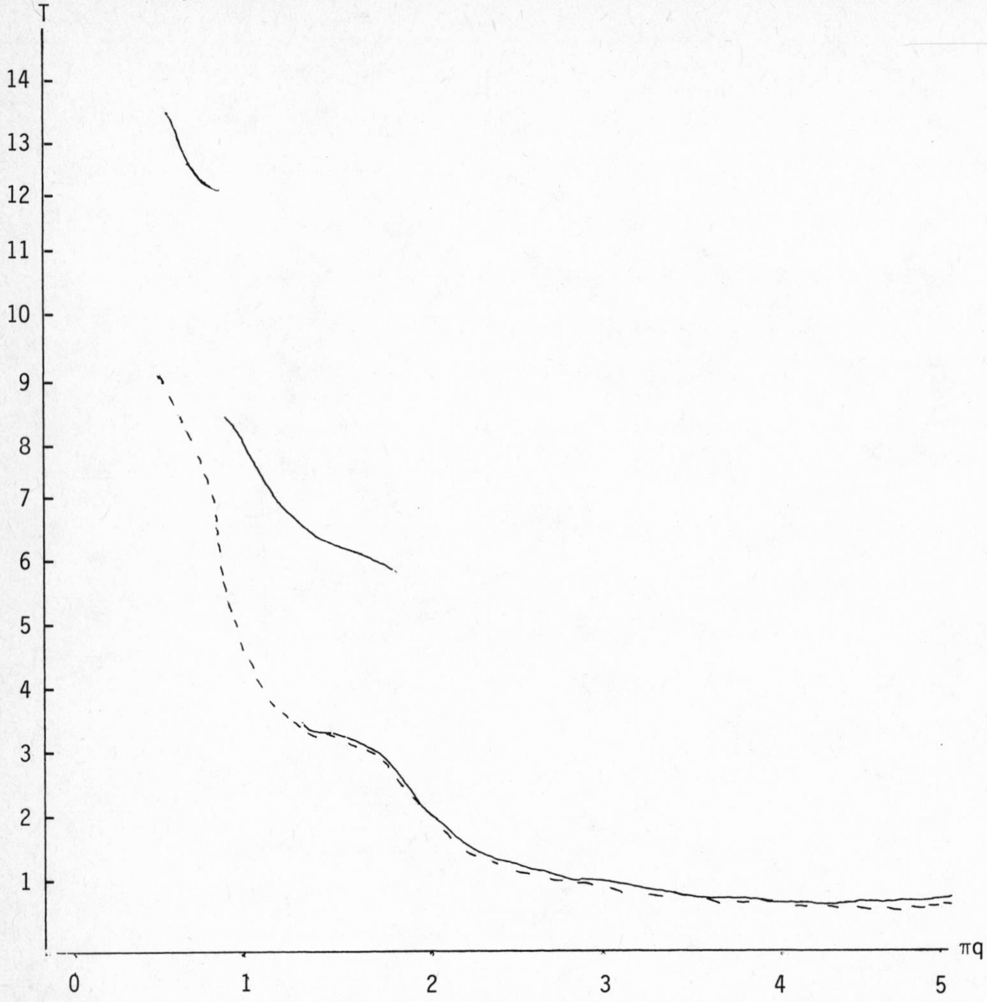

In the region under the full-line graph (which is considerably larger than the region under the broken-line graph - which corresponds to the control-unconstrained case -) there exists a unique optimal (feedback) solution for all initial states $x_o \in R^n$.

9.5 REFERENCES

[1] R.F. Brammer, Controllability in linear autonomous systems with positive controllers. *SIAM J. Control* Vol 10, pp 339-353, 1972.

[2] J. Stiglitz (ed.), *The Collected Scientific Papers of Paul A. Samuelson*. M.I.T. Press, Cambridge, Mass., 1966.

[3] S.J. Turnovsky, *Macroeconomic Analysis and Stabilization Policies*. Cambridge University Press, Cambridge, U.K. 1977.

[4] D.H. Martin and D.H. Jacobson, Optimal control laws for a class of constrained linear-quadratic problems. *Automatica*, Vol 15, pp 431-440, 1979.

[5] W.H. Fleming and R.W. Rishel, *Deterministic and Stochastic Optimal Control*. Springer-Verlag, New York, 1975.

[6] V.G. Boltyanskii, *Mathematical Methods of Optimal Control*. Holt, Rinehart, Winston, Publishers, New York, 1971.

[7] T.G. Hack, Extensions of the concept of regular synthesis as a sufficient condition for optimality. *SIAM J. Control* Vol 11, pp 358-374, 1973.

[8] M. Pachter, The linear-quadratic optimal control problem with positive controllers. To appear in *Int. J. Control*, 1980.

[9] J.V. Breakwell and Yu-Chi-Ho, On the conjugate point condition for the control problem. *Int. J. Engng. Sci.*, Vol 2, pp 565-579, 1965.

[10] I.M. Gelfand and S.V. Fomin, *Calculus of Variations*. Prentice-Hall, Englewood Cliffs, N.J., 1963.

[11] R.W. Brockett, *Finite-Dimensional Linear Systems*. Wiley, New York, 1970.

10. SPECIAL NON-LINEAR-QUADRATIC FORMULATIONS

10.1 INTRODUCTION

The deterministic linear-quadratic optimal control problem may be solved by a number of means. The problem is of interest both for this reason and because it plays an important role in control engineering applications.

In this lecture we show that there are certain non-linear-quadratic optimal control problems which can be solved by essentially the same techniques as those used for the linear-quadratic case. Though these problems do not arise as frequently in practice as does the linear-quadratic one, the results indicate the degree to which known results may be extended using conventional techniques.

10.2 INFINITE-TIME OPTIMAL CONTROL

It turns out to be extraordinarily difficult to solve finite-time non-linear-quadratic problems; this is essentially owing to the fact that a solution to the associated H-J-B equation is not immediate. The situ= ation is somewhat better in the infinite-time case, as a solution to the H-J-B equation can then be assumed to be independent of time, with con= sequent simplification.

We first consider the general infinite-time optimal control problem of minimizing with respect to $u(\cdot)$ the criterion

$$J(x_0,u(\cdot)) = \int_0^\infty L(x(t),u(t))dt \qquad (10.2.1)$$

subject to the dynamic system

$$\dot{x}(t) = f(x(t), u(t)), \quad x(0) = x_0 \qquad (10.2.2)$$

and the control constraint

$$u(t) \in \Omega \ , \ t \in [0,\infty), \qquad (10.2.3)$$

and the requirement that $x(t) \to 0$ as $t \to \infty$.

We make the following assumptions.

ASSUMPTION 10.2.1 The control functions allowed are those which be= long to U, where

$$U \triangleq \{u(\cdot) | u(\cdot) \text{ is piecewise continuous in } t \text{ and } u(t) \in \Omega \ , \ t \in [0,\infty)\}.$$

ASSUMPTION 10.2.2 The differential equation (10.2.2) has a solution defined on $[0,\infty)$ for each $u(\cdot) \in U$.

Our first theorem is an adaptation of Theorem 6.3.1 of Lecture 6 for the finite-time case.

THEOREM 10.2.3 [1] *Suppose there exists a continuously differentiable function* $V:\mathbb{R}^n \to \mathbb{R}^1$ *which is positive definite and which satisfies for all* $x \in \mathbb{R}^n$ *the steady-state* H-J-B *equation*

$$\min_{u \in \Omega} \ [L(x,u)+V_x(x)f(x,u)] = 0. \qquad (10.2.4)$$

Suppose also that all solutions of $\dot{x}(t) = f(x(t),u^0(x(t)))$ *go to zero as* $t \to \infty$ *and that along these solutions* $u^0(x(\cdot)) \in U$, *where*

$$u^0(x) \triangleq \arg\min_{u \in \Omega} \ [L(x,u)+V_x(x)f(x,u)]. \qquad (10.2.5)$$

Then, (10.2.1) *has a minimum in the class of control functions for which* $u(\cdot) \in U$ *and which cause* $x(t) \to 0$ *as* $t \to \infty$. *The minimum value of* J *is* $V(x_0)$ *and the minimizing controller is* $u^0(x)$.

PROOF The proof is quite similar to that of Theorem 6.3.1 and is given in full in [1] .

The next theorem establishes that a certain form can be assumed for $V_x(x)$, provided that a symmetry requirement is observed.

THEOREM 10.2.4 *The form* $V_x(x) \triangleq x'S(x)$, $S:\mathbb{R}^n \to \mathbb{R}^{n \times n}$ *is permitted if and only if the matrix* $\frac{\partial}{\partial x}[S'(x)x]$ *is symmetric.*

PROOF It is well known that $\phi: \mathbb{R}^n \to \mathbb{R}^n$ is the derivative of a scalar function $V(x)$, $V:R^n \to R^1$, if and only if $\frac{\partial}{\partial x}\phi(x)$ is symmetric.

We apply the above theorems to an illustrative example.

EXAMPLE 10.2.5 We consider the performance criterion

$$J(x_0,u(\cdot)) = \int_0^\infty [\tfrac{1}{8}(x'(t)Qx(t))^2 + \tfrac{1}{2}u^2(t)]\,dt \qquad (10.2.6)$$

with Q positive definite, and

$$\dot{x}(t) = Ax(t)u(t). \qquad (10.2.7)$$

The H-J-B equation is

$$\min_u \; [\tfrac{1}{8}(x'Qx)^2 + \tfrac{1}{2}u^2 + x'S(x)Axu] = 0. \qquad (10.2.8)$$

Assuming for convenience that $S(x)$ is symmetric, we see that (10.2.8) yields

$$u^0(x) = -\tfrac{1}{2}x'[S(x)A+A'S(x)]x \qquad (10.2.9)$$

and

$$\tfrac{1}{8}(x'Qx)^2 - \tfrac{1}{8}\{x'[S(x)A+A'S(x)]x\}^2 = 0. \qquad (10.2.10)$$

If we choose S to be a constant matrix which satisfies

$$Q \pm [SA+A'S] = 0 \qquad (10.2.11)$$

then (10.2.10) is satisfied. As Q is positive definite it follows (cf. Theorem 3.2.1) that if the real parts of the eigenvalues of A are all negative (positive) then $Q + (-)[SA+A'S] = 0$ yields a unique positive definite solution S^+ or S^-.

Accordingly, we have

$$V^+(x) = \tfrac{1}{2}x'S^+x, \quad u^0(x) = x'Qx \tag{10.2.12}$$

or

$$V^-(x) = \tfrac{1}{2}x'S^-x, \quad u^0(x) = -x'Qx. \tag{10.2.13}$$

It follows also from Lyapunov stability theory that (10.2.7) is asymp= totically stable in the large under the appropriate controllers (10.2.12) or (10.2.13). Whether A is a stability or an instability matrix, we have

$$\begin{aligned} \frac{d}{dt} V^{\pm}(x) &= \pm\, x'(S^{\pm}A+A'S^{\pm})x.x'Qx \\ &= -\,(x'Qx)^2, \end{aligned} \tag{10.2.14}$$

which is negative definite. As $V^{\pm}(x)$ is positive definite and radially unbounded, global asymptotic stability results. Incidentally, note that the condition required by Theorem 10.2.4 is satisfied.

10.3 SYSTEMS HOMOGENEOUS-IN-THE-INPUT

A system is homogeneous-in-the-input if it is of the form

$$\dot{x}(t) = \sum_{i=1}^{r} B_i(x(t))u_i(t), \quad x(0) = x_o. \tag{10.3.1}$$

We make the following assumption.

ASSUMPTION 10.3.1 The functions $B_i(x)$, $B_i : \mathbb{R}^n \to \mathbb{R}^n$, $i=1,\ldots,r$, are continuous.

In this section we investigate the behaviour of systems of the form (10.3.1), especially under the feedback control $u_i(x) = g_i(x)$, $i=1,\ldots,r$, where the $g_i(\cdot)$ are continuous functions of x.

10.3.1 Stability

It turns out that despite the presence of the nonlinear functions $B_i(x)$ in (10.3.1), an almost complete answer can be stated to the question of stabilizability of (10.3.1).

THEOREM 10.3.2 [1] *Suppose that* $u_i(x) = g_i(x)$, $i=1,\ldots,m$, *and that the* B_i *and* g_i *are once continuously differentiable in* x. *A necessary condition for* $x = 0$ *to be an asymptotically stable equilibrium point of* (10.3.1) *is that there exists a positive definite function* V, *defined and once continuously differentiable on an open neighbourhood* Ω *of* $x=0$, *such that there is no non-zero* $x \in \Omega$ *for which*

$$V_x(x)B_i(x), \quad i=1,\ldots,r \qquad (10.3.2)$$

are all zero.

An almost exactly similar theorem provides sufficient conditions for optimality.

THEOREM 10.3.3 [1] *A sufficient condition for* $x = 0$ *to be made into an asymptotically stable equilibrium point of* (10.3.1) *is that there exists a positive definite function* V, *defined and once continuously differentiable on some open neighbourhood* Ω *of* $x=0$, *such that there is no non-zero* $x \in \Omega$ *for which*

$$V_x(x)B_i(x), \quad i=1,\ldots,r \qquad (10.3.3)$$

are all zero. Accordingly, the feedback law $u_i(x) \triangleq V_x(x)B_i(x)$ *causes* $x = 0$ *to be an asymptotically stable equilibrium point and, if* V *is radially unbounded, the stability is global.*

PROOF Let us set $u_i(x) = -V_x(x)B_i(x)$, $i=1,\ldots,r$. Then the $u_i(x)$ are continuous and

$$\dot{V}(x) = -\sum_{i=1}^{r} [V_x(x)B_i(x)]^2 < 0, \quad x \neq 0. \tag{10.3.4}$$

Asymptotic stability, and global asymptotic stability when V is radially unbounded, follow from standard Lyapunov theory (see Lecture 3).

Theorem 10.3.3 is especially useful in those cases where (10.3.1) cannot be stabilized by means of constant controls $u_i = k_i$, $i=1,\ldots,r$, as the next example illustrates.

EXAMPLE 10.3.4 We consider the bilinear dynamic system

$$\dot{x}(t) = B_1x(t)u_1(t) + B_2x(t)u_2(t) \tag{10.3.5}$$

where

$$B_1 = \begin{bmatrix} 0 & 1 \\ 1 & 0 \end{bmatrix}, \quad B_2 = \begin{bmatrix} -1 & 0 \\ 0 & 1 \end{bmatrix}. \tag{10.3.6}$$

Clearly there are no constants k_1, k_2 for which $B_1k_1 + B_2k_2$ is a stability matrix. On the other hand this system can be stabilized by the controllers

$$\begin{aligned} u_1(x) &= -2x_1x_2 \\ u_2(x) &= x_1^2 - x_2^2 \end{aligned} \tag{10.3.7}$$

which correspond to the choice $V(x) = \frac{1}{2}x'x$.

10.3.2 Optimal control

We have shown that subject to very reasonable conditions (10.3.1) is stabilizable and that stabilizing controllers can be readily constructed. The next result yields stabilizing controllers which also optimize the performance criterion

$$J(x_o,u(\cdot)) = \int_0^\infty \{q(x(t)) + \frac{1}{2(p+1)} \cdot \sum_{i=1}^r u_i^{2(p+1)}(t)\}dt \tag{10.3.8}$$

where $q(x)$ is a positive definite function $x \in \mathbb{R}^n$, and p is a non-negative integer.

THEOREM 10.3.5 [1] *Suppose there exists a radially unbounded positive definite function* $V(x)$ *which is once continuously differentiable, which satisfies the H-J-B equation*

$$\min_{u_1,\ldots,u_r} [q(x) + \frac{1}{2(p+1)} \sum_{i=1}^r u_i^{2(p+1)} + \sum_{i=1}^r V_x(x)B_i(x)u_i] = 0. \tag{10.3.9}$$

Then, the controls

$$u_i(x) = -[V_x(x)B_i(x)]^{\frac{1}{2p+1}}, \quad i=1,\ldots,r \tag{10.3.10}$$

globally asymptotically stabilize (10.3.1) *and minimize* (10.3.8) *in the class of control functions which cause* $x(t) \to 0$ *as* $t \to \infty$.

PROOF Carrying out the minimization in (10.3.9) yields (10.3.10) and

$$q(x) - \frac{2p+1}{2(p+1)} \sum_{i=1}^m [V_x(x)B_i(x)]^{\frac{2(p+1)}{2p+1}} = 0. \tag{10.3.11}$$

Note that from (10.3.11) and the assumption on $q(x)$ it follows that

$V_x(x)B_i(x)$, $i=1,\ldots,r$ cannot all be zero for a non-zero $x \in \mathbb{R}^n$ and accordingly it is easy to show that (10.3.10) asymptotically stabilizes (10.3.1).

Note that because of the special form of the H-J-B equation we can, without loss of generality, replace q(x) in (10.3.8) by

$$\frac{2p+1}{2(p+1)} \sum_{i=1}^{r} [V_x(x)B_i(x)]^{\frac{2(p+1)}{2p+1}} \qquad (10.3.12)$$

so that instead of specifying q(x) and solving for V(x) we can choose a suitable V(x) to specify q(x).

EXAMPLE 10.3.6 Returning to Example 10.3.4 we note that (10.3.7) minimizes (10.3.8) with p = 0 and

$$q(x) = \sum_{i=1}^{2} [x'B_i x]^2 = \tfrac{1}{2}(x_1^2 + x_2^2)^2. \qquad (10.3.13)$$

10.3.3 Non-homogeneous extension: systems linear-in-the-input

Here we consider the equation

$$\dot{x}(t) = f(x(t)) + \sum_{i=1}^{r} B_i(x(t))u_i(t), \quad x(0) = x_o. \quad (10.3.14)$$

THEOREM 10.3.7 [1] *Suppose that there exists a radially unbounded, positive definite function* V(x) *which is once continuously differentiable such that* $V_x(x)f(x)$ *is negative semidefinite. Suppose further that there is no non-zero* $x \in \mathbb{R}^n$ *for which* $V_x(x)f(x)$ *and* $V_x(x)B_i(x)$, $i=1,\ldots,r$, *are all zero. Then,*

$$u_i(x) = -V_x(x)B_i(x), \quad i=1,\ldots,r \qquad (10.3.15)$$

globally asymptotically stabilizes (10.3.1)*and minimizes*

$$J(x_o,u(\cdot)) = \int_0^\infty \{q(x(t)) + \tfrac{1}{2} \sum_{i=1}^{r} u_i^2(t)\}dt \qquad (10.3.16)$$

in the class of control functions which cause $x(t) \to 0$ *as* $t \to \infty$. *Here,* $q(x)$ *is, without loss of generality, given by the positive definite expression*

$$-V_x(x)f(x) + \tfrac{1}{2} \sum_{i=1}^{r} [V_x(x)B_i(x)]^2. \qquad (10.3.17)$$

10.4 QUADRATIC SYSTEMS

We consider here the system

$$\frac{d}{dt} f_i(x(t)) = \tfrac{1}{2}x'(t)A_i x(t) + x'(t)B_i u(t), \quad i=1,\dots,n, x(0) = x_o, \qquad (10.4.1)$$

where the f_i are continuous functions of x. We assume that (10.4.1) has a solution for each piecewise continuous control function.

The performance criterion which we choose to minimize is

$$J(x_o,u(\cdot)) = \int_0^\infty \tfrac{1}{2}(x'(t)Qx(t) + u'(t)Ru(t))dt \qquad (10.4.2)$$

where Q and R are positive definite symmetric matrices. Our result is as follows.

THEOREM 10.4.1 *Suppose that there exists a constant vector* $c \in \mathbb{R}^n$ *which satisfies the Riccati-like equation*

$$Q + \sum_{i=1}^{n} c_i A_i - \Big(\sum_{i=1}^{n} c_i B_i\Big)R^{-1}\Big(\sum_{i=1}^{n} c_i B_i\Big)' = 0, \qquad (10.4.3)$$

and that each solution of the differential equations

$$\frac{d}{dt} f_i(x(t)) = \tfrac{1}{2}x'(t)A_i x(t) - x'(t)B_i R^{-1}\Big(\sum_{i=1}^{n} c_i B_i\Big)x(t) \qquad (10.4.4)$$

$$x_i(0) = x_{oi}, \quad i=1,\dots,n$$

is defined for $t \in [0,\infty)$ *and that* $x(t) \to 0$ *as* $t \to \infty$. *Then, the minimum value of* (10.4.2) *in the class of controls which cause* $x(t) \to 0$ *as* $t \to \infty$ *is*

$$J = c'[f(x_0)-f(0)] \tag{10.4.5}$$

and an optimal controller which achieves this minimum is

$$u(x) = -R^{-1}\Big(\sum_{i=1}^{n} c_i B_i\Big)x. \tag{10.4.6}$$

PROOF Along a solution of (10.4.1) we have

$$c'f(x_0) - c'f(x(t)) + \int_0^t \Big[\tfrac{1}{2}x'(t)\sum_{i=1}^{n} c_i A_i x(\tau) + x'(t)\sum_{i=1}^{n} c_i B_i u_i(\tau)\Big]d\tau = 0. \tag{10.4.7}$$

Adding this to (10.4.2) and letting $t \to \infty$ we have

$$\begin{aligned} J(x_0,u(\cdot)) = {} & c'f(x_0) - \lim_{t\to\infty} c'f(x(t)) \\ & + \int_0^\infty \tfrac{1}{2}\Big\{x'(t)\Big[Q + \sum_{i=1}^{n} c_i A_i - \Big(\sum_{i=1}^{n} c_i B_i\Big)R^{-1}\Big(\sum_{i=1}^{n} c_i B_i\Big)'\Big]x(t) \\ & + \Big[u(t)+R^{-1}\Big(\sum_{i=1}^{n} c_i B_i\Big)'x(t)\Big]'R\Big[u(t)+R^{-1}\Big(\sum_{i=1}^{n} c_i B_i\Big)'x(t)\Big]\Big\}dt. \end{aligned} \tag{10.4.8}$$

As we allow only those controls which cause $x(t) \to 0$ as $t \to \infty$, we see that by the continuity of f, $\lim_{t\to\infty} c'f(x(t))$ is equal to $c'f(0)$. Now by assumption, (10.4.6) acting on (10.4.1) causes $x(t) \to 0$ as $t \to \infty$, so by inspection (10.4.6) minimizes J and the minimum value is given by (10.4.5).

The above result appears to be of only academic interest, as (10.4.3) is not easy to solve (it has a solution only for special choices of Q, A_i, B_i and R) and it is not easy to check that solutions of (10.4.4) exist and go to zero as $t \to \infty$.

10.5 REFERENCE

[1] D.H. Jacobson, *Extensions of linear-quadratic control, optimization and matrix theory*. Academic Press, New York, 1977.

11. HYBRID CRITERIA AND PARAMETER SENSITIVITY

11.1 INTRODUCTION

This lecture is devoted to the study of discrete-time control systems.

The discretization scheme introduced and analyzed here, and which yields the discrete-time control system, is a natural discretization scheme if one is constrained to measure (or sample) the state in the linear-quadratic continuous-time optimal control problem on the prescribed discrete-time set $0,T,2T,\dots,NT$, with $T = \tilde{T}/N$. As, however, this scheme differs to some extent from the discretization scheme which is usually discussed in the literature [1], we first, for the sake of completeness, introduce the standard discretization scheme, as follows.

Consider the continuous-time control system

$$\dot{x} = Ax + Bu, \quad x(0) = x_0, \quad 0 \leqslant t \leqslant \tilde{T}, \quad \tilde{T} > 0 \tag{11.1.1}$$

and cost functional

$$J(u,x_0) = \int_0^{\tilde{T}} [u'(t)u(t) + x'(t)Qx(t)]\,dt. \tag{11.1.2}$$

Here $x,x_0 \in \mathbb{R}^n$, $u \in \mathbb{R}^m$ and the real matrices A,B and Q have the appro= priate dimensions. In addition, the matrix Q is symmetric.

In a sampled-date system the continuous system (11.1.1) is driven by an input specified at discrete time points Ti, $i=0,1,2,\dots,N$ and has state-variable information only at these discrete time points. Speci= fically, it is assumed that the control $u(t)$ is then a piecewise con= stant function of time, i.e. $u(t) = u(Ti) = u_i$, $Ti \leqslant t \leqslant T(i+1)$, and the state is sampled at these discrete time points. In this case

$$x_{i+1} = \tilde{A}\, x_i + \tilde{B}u_i \tag{11.1.3}$$

and the corresponding discrete cost functional is

$$J(u,x_o) = \sum_{i=1}^{N} x_i' \tilde{Q} x_i + 2x_i' Mu_i + u_i' \tilde{R} u_i, \qquad (11.1.4)$$

where

$$\tilde{A} \triangleq e^{AT}, \ \tilde{B} \triangleq \int_0^T e^{At}Bdt, \quad \tilde{Q} \triangleq \int_0^T e^{A't}Qe^{At}dt,$$

$$M \triangleq \int_0^T e^{A't} Q D(t)dt \quad \text{and } \tilde{R} \triangleq T + \int_0^T D'(t) Q D(t)dt, \text{where } D(t) \triangleq \int_0^t e^{At}Bdt.$$

When this discretization scheme is used, two problems unique to the discrete-time case must be considered, viz.

(i) the possibility of a singular control effort weighting matrix (in the continuous-time original problem)[1];

(ii) loss of controllability owing to sampling [2].

This discretization scheme is discussed in [1], while [3] contains an analysis of the sensitivity of the solution to the sampling parameter T.

In this lecture we assume that the system state is available only in the discrete-time set Ti, i=0,1,...,N and we discretize the continuous-time optimal control problem (11.1.1), (11.1.2) by writing the trajec= tory-dependent term in the performance functional (11.1.2) in the form of a Riemann sum on the time partition [0,T,...,NT]; leaving the system dynamics, however, in continuous form. Specifically, in Section 11.2 we apply this discretization scheme to the solution of hybrid linear-quadratic optimal control problems, while in Section 11.3 we then devote attention to the question of feedback implementation.

In Section 11.4 we discuss the differentiability properties of the solution to the algebraic Riccati equation which arises in Section 11.3.

In Section 11.5 we consider continuous-discrete gain transformations (i.e. the sensitivity to the sampling parameter T), and in Section 11.6 we comment on the information pattern/performance trade-off.

The ensuing method of solution for the (hybrid) optimal control problem in Section 11.7 hinges upon the efficient and fast algorithms employed in solving block-tridiagonal systems of linear equations which arise in the study of partial differential equations.

Throughout, attention will be given to the question of the existence of solutions and the efficient verification of such existence.

11.2 A HYBRID LINEAR-QUADRATIC OPTIMAL CONTROL PROBLEM

Consider the linear time-invariant system (11.1.1) with cost functional (11.1.2). It is well known (see Lecture 4) that the solution to the optimal control problem $\min_u(J(u,x_0))$ is given by $u^*(t) = -B'P(\tilde{T}-t)x(t)$ and $J(u^*,x_0) = x_0'P(\tilde{T})x_0$, where $P(\tau)$, $0 \leqslant \tau \leqslant \tilde{T}$ is the solution (if it exists) to the matrix Riccati differential equation

$$\dot{P} = A'P + PA - PBB'P + Q, \; P(0) = 0.$$

Note that the optimal control is in feedback form; it is however neces= sary, by solving the Riccati differential equation (possibly on a digital computer), to precompute $P(\tau)$ and store its values for all $0 \leqslant \tau \leqslant \tilde{T}$, or to solve the Riccati differential equation backwards in time. One would obviously solve for/store $P(\tau)$ on a discrete-time set. Alternatively, the optimal control $u^*(t)$ could be obtained in open-loop form, provided the solution to a linear 2n-dimensional TPBVP was calculated. This is again computationally equivalent to solving the Riccati differential equation and storing the discretized solution.

We approach the above optimal control problem by discretizing in the following manner. Let N be a positive integer, let $T = \tilde{T}/N$ and replace $\int_0^{\tilde{T}} x'(t)Qx(t)dt$ in the performance functional by the Riemann sum $T \sum_{i=1}^{N} x'(Ti)Qx(Ti)$. We then have the 'hybrid' optimal control problem

$$\min_u \tilde{J}(u,x_0) \triangleq \min_{u(\cdot)} \left[\int_0^{TN} u'(t)u(t)dt + T \sum_{i=1}^{N} x_i' Qx_i\right] \tag{11.2.1}$$

subject to

$$\dot{x} = Ax + Bu,\ x(o) = x_0,\ 0 \leqslant t \leqslant TN \text{ and } x_i \triangleq x(Ti),\ i=1,2,\ldots,N. \tag{11.2.2}$$

Note that from the implementation point of view this is a natural dis= cretization scheme to which to have recourse if one is constrained to measure (or sample) the state in the linear-quadratic optimal control problem (11.1.1), (11.1.2), at the prescribed discrete time set Ti, $i=0,1,\ldots,N$.

It is now evident that in order to solve this hybrid optimal control problem, one should choose a 'trajectory' $(x_1,x_2,\ldots,x_N)$ such that if we denote by $E(x_{i-1},x_i)$ the minimal 'control energy' $\int_{T(i-1)}^{Ti} u'(t)u(t)dt$ required to perform the transfer from x_{i-1} to x_i along the control sys= tem trajectories, then $\sum_{i=1}^{N} [E(x_{i-1},x_i) + Tx_i'Qx_i]$ is minimal. The minimum energy control which steers x_{i-1} to x_i, i.e. our optimal control, and the corresponding minimum control energy are (see for example [4]) given by

$$u^*(t) = -B'e^{-A'[t-(i-1)T]}\xi_i,\ (i-1)\,T \leqslant t \leqslant iT,\ i=1,2,\ldots,N \tag{11.2.3}$$

and

$$E(x_{i-1},x_i) = \xi_i'W\xi_i, \text{ respectively, } i=1,2,\ldots,N$$

where

$$W(t) \triangleq \int_0^t e^{-A\tau}BB'e^{-A'\tau}d\tau \quad \text{and} \quad W \triangleq W(T)$$

and ξ_i is such that

$$W\xi_i \triangleq (x_{i-1}-e^{-AT}x_i).$$

We have thus transformed the hybrid optimal control problem into the discrete optimal control problem

$$\min_{\xi_i, i=1,2,\ldots,N} \sum_{i=1}^{N} \xi_i'W\xi_i + Tx_i'Qx_i \tag{11.2.4}$$

subject to

$$x_i = e^{AT}x_{i-1}-e^{AT}W\xi_i, \quad x_o \text{ given and } i=1,2,\ldots,N. \tag{11.2.5}$$

It is interesting to observe that we are now faced with a discrete op= timal control problem which is somewhat different from the discrete optimal control problems (11.1.3), (11.1.4) which arise when continuous-time systems, in sample data control systems, are approximated by dis= crete systems, or when the physical system in question is inherently discrete (see, for example,[1] and [5]). In particular, it is to be noted that controllability of the continuous system implies by Theorem 1.3.2 that the discrete system control matrix $e^{AT}W$ is an $n \times n$ non-singular matrix. In addition, the above discrete control system is controllable iff the original continuous control system is controllable (cf. (ii) in Section 11.1); also, the discrete optimal control problem is non-singu= lar iff the original continuous control system is controllable (cf. (i) in Section 11.1).

11.3 FEEDBACK IMPLEMENTATION

In the rest of this lecture we shall assume that the continuous system is controllable and observable. Let n be an integer, $1 \leqslant n \leqslant N+1$, and denote by $V_n(x_{n-1})$ the minimum value of the performance criterion along trajectories of the discrete system, on the time set $T(n-1), Tn, \ldots, TN$; the trajectories emanating from x_{n-1}. Obviously, $\tilde{J}(u^*, x_{n-1}) = V_n(x_{n-1})$. Moreover, by a standard dynamic programming argument, given in the final section of Lecture 4, we have,

$$V_n(x_{n-1}) = x'_{n-1} P_n x_{n-1},$$

where P_n, $n = N, N-1, \ldots, 1$, is the solution to the discrete Riccati equation which corresponds to the discrete system $(e^{AT}, -e^{AT}W, W, TQ)$, viz.

$$P_n = \tilde{A}'(P_{n+1}+TQ)\tilde{A} - \tilde{A}'(P_{n+1}+TQ)\tilde{A}[W^{-1}+\tilde{A}'(P_{n+1}+TQ)\tilde{A}]^{-1}\tilde{A}'(P_{n+1}+TQ)\tilde{A},$$

$$P_{N+1} = 0, \qquad (11.3.1)$$

We recall that here $\tilde{A} \triangleq e^{AT}$ and assume that

$$W^{-1} + \tilde{A}'(P_{n+1} + TQ)\tilde{A} > 0 \text{ for } n = N, N-1, \ldots, 1; \qquad (11.3.2)$$

also

$$\xi_n^* = [W + W\tilde{A}'(P_{n+1}+TQ)\tilde{A}W]^{-1} W\tilde{A}'(P_{n+1}+TQ)\tilde{A}x_{n-1}.$$

Equation 11.3.1 (and 11.3.2) can be simplified if we employ a matrix inversion lemma and if we define

$$\tilde{P}_n \triangleq P_n + TQ. \qquad (11.3.3)$$

We then obtain

$$\left.\begin{aligned} \tilde{P}_n &= TQ + \tilde{A}'\tilde{P}_{n+1}\tilde{A}(\tilde{A}'\tilde{P}_{n+1}\tilde{A}+W^{-1})^{-1}W^{-1} \\ &\text{or} \\ \tilde{P}_n &= TQ + W^{-1}(\tilde{A}'\tilde{P}_{n+1}\tilde{A}+W^{-1})^{-1}\tilde{A}'\tilde{P}_{n+1}\tilde{A} \end{aligned}\right\}, \tilde{P}_{N+1} = TQ \qquad (11.3.4)$$

$$n = N, N-1, \ldots, 1$$

and

$$\xi_n^* = (\tilde{P}_n - TQ)x_{n-1}. \qquad (11.3.5)$$

If in addition the matrices $\tilde{P}_N, \tilde{P}_{N-1},\ldots,\tilde{P}_1$ are invertible, then the Riccati equation (11.3.4) can be written in a 'symmetric' form upon reapplying a 'matrix inversion' type lemma. Then

$$\tilde{P}_n = TQ + (W + \tilde{A}^{-1}\tilde{P}_{n+1}^{-1}\tilde{A}'^{-1})^{-1}, \; n = N-1,N-2,\ldots,1, \qquad (11.3.6)$$

and in view of (11.3.4)

$$\tilde{P}_N = T\,[\,Q + \tilde{A}'Q\tilde{A}(T\tilde{A}'Q\tilde{A} + W^{-1})^{-1}W^{-1}]\,.$$

Note that if Q is positive semidefinite, i.e. $Q = C'C$ and the pair (A,C) is observable, then $\tilde{P}_k > 0$ for $k = N-n+1, N-n,\ldots,1$ and P_{N-n+2} must be computed from (11.3.4). In fact, we can summarize with the following theorem.

THEOREM 11.3.1 *Consider the optimal control problem* (11.2.4), (11.2.5) *and assume that the condition* (11.3.2) *holds (i.e. that we do not have the analogy of a finite 'escape time') where* $\tilde{P}_n$, *as defined in* (11.3.3) *is the solution to* (11.3.4). *We then have the following feedback imple= mentation* :

Solve (11.3.4) (or (11.3.6)) *for* $\tilde{P}_n$ *and then compute* ξ_n^* *from* (11.3.5). *In the time interval* $[T(n-1), Tn]$ *apply the optimal control given by* (11.2.3). □

Note that the ensuing control function is not necessarily continuous.

The feedback implementation is particularly simple if $\tilde{T} = \infty$ (i.e. $N = \infty$), for then the adjoint system is initialized (at the sampling rate) by the state feedback $(\tilde{P} - TQ)x_{n-1}$, where $\tilde{P}$ is the solution to the discrete algebraic Riccati equation

$$\tilde{P} = TQ + \tilde{A}'\tilde{P}\tilde{A}(W\tilde{A}'\tilde{P}\tilde{A} + I)^{-1} \qquad (11.3.7)$$

or

$$\tilde{P} = TQ + (W + \tilde{A}^{-1}\tilde{P}^{-1}\tilde{A}^{-1})^{-1} \qquad (11.3.7')$$

Note that W^{-1} does not appear in the equations (11.3.7) or (11.3.7').

In particular, we have

THEOREM 11.3.2 *Assume that* $Q \geqslant 0$. *If the continuous system is con= trollable (as we have assumed throughout) then there exists a solution* $\tilde{P} \geqslant 0$ *to the discrete algebraic Riccati equation* (11.3.7) *and for any sampling rate* $T > 0$ *there also exists a solution to the hybrid linear-quadratic infinite-time optimal control problem* (11.2.4), (11.2.5); *moreover, even if the continuous system is not controllable but there exists a solution to the continuous linear-quadratic infinite-time opti= mal control problem* (11.1.1), (11.1.2), *then for all* $T > 0$ *there also exists a solution* $\tilde{P} \geqslant 0$ *to the discrete algebraic Riccati equation* (11.3.7)*and a solution to the hybrid linear-quadratic infinite-time optimal control problem* (11.2.4), (11.2.5). □

11.4 THE DIFFERENTIABILITY PROPERTIES OF THE SOLUTION

We shall show that the solution $\tilde{P}(T)$ to equation (11.3.7') has the re= quired continuous differentiability properties with respect to the sampling parameter T and that therefore it is possible to apply Taylor's theorem (in Section 11.5) in order to approximate the solution $\tilde{P}$ to equation (11.3.7') near $T = 0$ by

$$\tilde{P} = P_0 + T\,\frac{\partial \tilde{P}}{\partial T}(0) + \frac{T^2}{2}\,\frac{\partial^2 \tilde{P}}{\partial T^2}(0) + O(T^3); \qquad (11.4.1)$$

in other words, we then have a solution of the discrete algebraic Riccati equation (11.3.7') which is of high order sensitivity in the parameter T.

Indeed, upon multiplying equation (11.3.7') from the right by $\tilde{A}'\tilde{P}\tilde{A}$ and from the left by $(\tilde{P}-TQ)$, we obtain the quadratic equation in $\tilde{P}$

$$\tilde{A}'\tilde{P}\tilde{A} - \tilde{P} - \tilde{P}W\tilde{A}'\tilde{P}\tilde{A} + TQ + TQW\tilde{A}'\tilde{P}\tilde{A} = 0. \qquad (11.4.2)$$

Next, we divide eqn. (11.4.2) by T and define the functions

$$\tilde{f}(\tilde{P},T)\colon \mathbb{R}^{n^2} \times (0,\infty) \to \mathbb{R}^{n^2}$$

$$\tilde{f}(\tilde{P},T) \triangleq \frac{\tilde{A}'\tilde{P}\tilde{A} - \tilde{P} - \tilde{P}W\tilde{A}'\tilde{P}\tilde{A}}{T} + Q + QW\tilde{A}'\tilde{P}\tilde{A},$$

and

$$f(\tilde{P},T) \triangleq \begin{cases} \tilde{f}(\tilde{P},T) & \text{on} \quad \mathbb{R}^{n^2} \times (0,\infty) \\ \lim\limits_{T \to 0} \tilde{f}(\tilde{P},T) & \text{on the } \mathbb{R}^{n^2} \text{ plane.} \end{cases}$$

Treating $\tilde{P}$ as a parameter, we expand the following expressions in Taylor formulae in T (which is permissible along with expansions in Taylor series - because $\tilde{A}$ and W are analytic in T):

$$\tilde{A}'\tilde{P}\tilde{A} = \tilde{P} + T(A'\tilde{P}+\tilde{P}A) + \frac{T^2}{2}(A'^2\tilde{P}+\tilde{P}A^2+2A'\tilde{P}A)$$
$$+ \frac{T^3}{6}(\tilde{P}A^3+3A'\tilde{P}A^2+3A'^2\tilde{P}A+A'^3\tilde{P}) + O(T^4),$$

and

$$W\tilde{A}\tilde{P}\tilde{A} = TBB'\tilde{P} + \frac{T^2}{2}(2BB'A'\tilde{P} + 2BB'\tilde{P}A - ABB'\tilde{P} - BB'A'\tilde{P}) +$$
$$+ \frac{T^3}{6}(BB'A'^2\tilde{P} + 2ABB'A'\tilde{P} + A^2BB'\tilde{P} + 3BB'A'^2\tilde{P} + 3BB'\tilde{P}A^2 +$$
$$+ 6\,BB'A'\tilde{P}A - 3ABB'A'\tilde{P} - 3BB'A'^2\tilde{P} - 3ABB'\tilde{P}A - 3BB'A'\tilde{P}A) + O(T^4).$$

Hence

$$f(\tilde{P},T) = A'\tilde{P} + \tilde{P}A - \tilde{P}BB'\tilde{P} + Q + \frac{T}{2}(A'^2\tilde{P} + \tilde{P}A^2 + 2A'\tilde{P}A - 2\tilde{P}BB'A'\tilde{P}$$

$$- 2\tilde{P}BB'\tilde{P}A + \tilde{P}ABB'\tilde{P} + \tilde{P}(BB'A'\tilde{P} + 2QBB'\tilde{P})) + \frac{T^2}{6}(\tilde{P}A^3 +$$

$$+ 3A'\tilde{P}A^2 + 3A'^2\tilde{P}A + A'^3\tilde{P} - \tilde{P}\,BB'A'^2\tilde{P} - 2\tilde{P}ABB'A'\tilde{P} -$$

$$- \tilde{P}\,A^2BB'\tilde{P} + 3\tilde{P}\,BB'A'^2\tilde{P} - 3\tilde{P}\,BB'\tilde{P}A^2 - 6\,\tilde{P}BB'A'\tilde{P}A +$$

$$+ 3\,\tilde{P}\,ABB'A'\tilde{P} + 3\tilde{P}\,BB'A'^2\tilde{P} + 3\,\tilde{P}\,A\,BB'\tilde{P}A + 3\tilde{P}\,BB'A'\tilde{P}A +$$

$$+ 6\,Q\,BB'A'\tilde{P} + 6\,Q\,BB'\tilde{P}A - 3\,Q\,A\,BB'\tilde{P} - 3\,Q\,BB'A'\tilde{P}) + O(T^4). \tag{11.4.3}$$

From (11.4.3) we conclude the following

(i) The function $f \in C^2$ on $\mathbb{R}^{n^2} \times [0,\tau)$ for $\tau > 0$ sufficiently small.

(ii) Letting P be the unique real symmetric positive definite solution to the continuous algebraic Riccati equation

$$A'P + PA + PBB'P + Q = 0 \tag{11.4.4}$$

we note that

$$f(P,0) = 0.$$

(iii) $$f_{\tilde{P}}(P,0) \triangleq \frac{\partial f}{\partial \tilde{P}}(P,0) = 2(A-BB'P)$$

and since the continuous optimal closed-loop system is asymptotically stable, i.e. $Re(\lambda(A-BB'P)) < 0$,

$$\det(f_{\tilde{P}}(P,0)) = \det(2(A-BB'P)) \neq 0.$$

Hence, the implicit function Theorem 7-6 from [6] is applicable, and we conclude that for small $T > 0$ the unique function $\tilde{P}(T)$ (which is determined by $f(\tilde{P}(T),T) = 0$, i.e. by eq. (11.4.2)) is C^1 and $\tilde{P}(0) = P$.

Moreover, since P is positive definite we conclude that, provided $0 \leqslant T < \tau$, τ sufficiently small, $\tilde{P}(T)$ is indeed the unique real symme= tric positive definite solution of equation (11.4.2), and hence of (11.3.7'), as ascertained by Theorem 11.3.2.

Finally, note that $\frac{d\tilde{P}(T)}{dT}$ is explicitly given by

$$\frac{d\tilde{P}}{dT}(T) = -\left(f_P(\tilde{P}(T),T)\right)^{-1} f_T(\tilde{P}(T),T), \qquad (11.4.5)$$

and since in view of (11.4.3), $f \in C^2$, we conclude from (11.4.5) that also $\tilde{P}(T) \in C^2$ for $T \geqslant 0$, T sufficiently small. In fact, $\frac{d^2\tilde{P}}{dT^2}$ is given recursively by

$$\frac{d^2\tilde{P}}{dT^2}(T) = -\left(f_{\tilde{P}}(\tilde{P}(T),T)\right)^{-1} \Big[f_{\tilde{P}\tilde{P}}(\tilde{P}(T),T)\left(\frac{d\tilde{P}(T)}{dT}\right)^2 + 2 f_{\tilde{P}T}(\tilde{P}(T),T)\,\frac{d\tilde{P}(T)}{dT} + f_{TT}(\tilde{P}(T),T)\Big].$$

The argument extends easily to higher derivatives. The reader is also referred to [7] where a similar analysis is performed for the continuous-time algebraic matrix Riccati equation.

11.5 TRANSFORMATIONS BETWEEN THE CONTINUOUS-TIME FEEDBACK GAIN AND THE HYBRID FEEDBACK GAIN

In the hybrid regulator (i.e. $\tilde{T} = \infty$, or $N = \infty$) the digital-to-analog converter (the adjoint system) is initialized, at the sampling rate, by the discrete-time state feedback $(\tilde{P}-TQ)x_{n-1}$, where $\tilde{P}$ is the solution to the discrete algebraic Riccati equation (11.3.7) or (11.3.7'). In continuous-time linear-quadratic synthesis, a great deal of attention has been devoted to the derivation of the continuous-time optimal gain, determined by the solution P to the continuous-time algebraic Riccati equation (11.4.4). In particular, it has been shown that there is a simple solution of the continuous-time algebraic Riccati equation if the original continuous-time system (A,B,C) (here $Q = C'C$) is controllable and observable, as we are assuming. What is needed then is a simple transformation between P and $\tilde{P}$.

To achieve this, we first point out the following.

(a) In view of Section 11.4 the matrix $\tilde{P}$ can, for small T, be expanded using Taylor's formula:

$$\tilde{P} = P_0 + TP_1 + T^2P_2 + T^3P_3 + O(T^4).$$

(b) If X,Y,Z are n×n matrices, then

$$(I + TX + T^2Y + T^3Z + \ldots)^{-1} = I - TX + T^2(X^2-Y) + T^3(XY + YX - Z - X^3) + \ldots$$

(c) $W = TBB' - \frac{1}{2} T^2(ABB' + BB'A') + \frac{1}{6} T^3(BB'A'^2 + 2ABB'A' + A^2BB')+\ldots$ (11.5.1)

$$\tilde{A}'\tilde{P}\tilde{A} = P_0 + T(P_1 + A'P_0 + P_0A) + \tfrac{1}{2}T^2(2P_2 + 2A'P_1 + A'^2P_0 + 2P_1A + 2A'P_0A + P_0A^2) + \tfrac{1}{6} T^3(6 P_3 + 6A'^2P_2 + 3 A'^2P_1 + A'^3P_0 + 6 P_2 A + 6 A'P_1A + 3 A'^2P_0A + P_0A^3 + 3P_1A^2 + 3 A'P_0A) + O(T^4). \quad (11.5.2)$$

(d) (A,B,C) a minimal realization implies that the 'closed-loop' system matrix

$$\bar{A} \triangleq A-BB'P$$

is asymptotically stable, so that the Lyapunov equation associated with $\bar{A}$ has a unique solution.

Then, by substituting (11.5.1), (11.5.2) into (11.3.7) and in view of (b) and (d) we conclude, after some tedious computations, that

$$\begin{cases} P_0 = P \\ P_1 = \frac{1}{2}Q, \end{cases} \qquad (11.5.3)$$

and

P_2 is the unique symmetric solution to the linear Lyapunov equation

$$\bar{A}'P_2 + P_2\bar{A} = M \qquad (11.5.4)$$

where the symmetric matrix M is given by

$$M \triangleq \frac{1}{4} QA\bar{A} - \frac{1}{4}\bar{A}'A'Q - \frac{1}{6} PA^2\bar{A} - \frac{1}{6} \bar{A}'A'^2P - \frac{1}{2}A'PBB'P\bar{A} +$$

$$+ \frac{1}{3} PABB'A'P - \frac{1}{4} PBB'PBB'Q - \frac{1}{4} A'PBB'Q -$$

$$- \frac{1}{2} PBB'PBB'A'P + \frac{1}{4} QBB'A'P.$$

Also, it turns out that if we take $\tilde{P} = P_0 = P$, then $\tilde{P}$ is a solution to the discrete Riccati equation (11.3.7'), with residual error $O(T^2)$; if we take $\tilde{P} = P + \frac{1}{2}TQ$, then $\tilde{P}$ is a solution to the discrete Riccati equa= tion with residual error $O(T^3)$; and if we take $\tilde{P} = P - \frac{1}{2}TQ + T^2P_2$, then $\tilde{P}$ is a solution to the discrete Riccati equation with residual error $O(T^4)$; etc. In addition, since the hybrid cost is $x_0'(\tilde{P}-TQ)x_0$, the continuous cost is $x_0'Pxo$, and $\tilde{P} = P + \frac{1}{2}TQ + \ldots$

We thus conclude that the discretization error is of the order $\frac{1}{2}TQ$.

Finally, the reader is referred to [3] for continuous-discrete feedback transformations in the standard discretization scheme presented in Section 6.1; there the coefficients $P_1, P_2, \ldots$ are different.

11.6 THE PERFORMANCE TRADE-OFF

If the information constraint 'x(t) is available only on the discrete time set t = Ti,i=0,1,...'(i.e. sampling is imposed upon us),then a plausible suboptimal procedure is to use the discretized performance criterion (11.2.1), and we then have the hybrid optimal control problem (11.2.1), (11.2.2). Fortunately, we now have a situation where it is possible to assess the degradation in performance as a result of the suboptimal control strategy. Specifically, we apply to the control system (11.2.2) the control

$$u(t) = -B'e^{-A'[t-(j-1)T]}\xi_j^*, \quad T(j-1) \leqslant t < Tj, \quad j=1,2,\ldots \qquad (11.6.1)$$

and calculate

$$J(u,x_0) = \int_0^\infty (x'Qx + u'u)dt; \qquad (11.6.2)$$

here ξ_j^* is determined by (11.3.5) and (11.3.7').

Indeed, the following has been shown [8] to hold.

THEOREM 11.6.1 Upon applying the hybrid optimal control (11.6.1) *(see also* (11.3.5) *and* (11.3.7')) *to the system* (11.1.1), *the value of the performance criterion* (11.6.2) *is* $x_0'Kx_0$, *where* K *is the unique real symmetric and positive-definite solution to the discrete Lyapunov equation*

$$[I-(\tilde{P}-TQ)W]e^{A'T}Ke^{AT}[I-W(\tilde{P}-TQ)]-K = -[E+(\tilde{P}-TQ)W(\tilde{P}-TQ)] \qquad (11.6.3)$$

where

$$E \triangleq \int_0^T [I-(\tilde{P}-TQ)W(t)]e^{A'T}Q\ell^{AT}[I-W(t)(\tilde{P}-TQ)]dt.$$

□

Note that by definition the performance degradation is $x_0'(K-P)x_0$. The difference in cost $x_0'(K-P)x_0 \geqslant 0$ can be interpreted as the relative value of the continuous optimal control over the suboptimal control determined upon discretization of the performance functional; alterna= tively, if we denote by K(T) the unique positive-definite solution to the Lyapunov equation (11.6.3) associated with the sampling rate T, then if $T_1 > T_2$, we have that

$$x_0'(K(T_1)-K(T_2))x_0$$

is the amount we would be willing to pay for decreasing the sampling interval from T_1 to T_2.

Finally, by performing a sensitivity analysis similar to the analysis in Sections 11.4 and 11.5, the following statement may be deduced.

THEOREM 11.6.2 [8] *The Taylor expansion of* K(T) (near T = 0) *is*

$$K(T) = P + T^2 K_2 + O(T^3),$$

i.e.

$$\left. \frac{\partial K(T)}{\partial T} \right|_{T = 0} = 0. \qquad \square$$

Hence for small T (i.e. a small discretization interval) the performance criterion is insensitive to the interval (we have nil sensitivity) and the value of the performance index (11.6.2) is very close to the classical linear-quadratic regulator value of $x_0'Px_0$; this is thus a good recommendation for employing the discretization procedure which we have described.

11.7 THE OPEN-LOOP SOLUTION

In the following example one is given from the outset a hybrid optimal control problem where the open-loop optimal control solution is of interest. Specifically, consider the situation of a statically loaded elastic string:

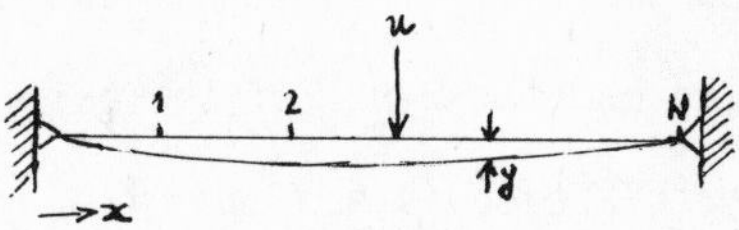

Here u is the load per unit length and y is the vertical displacement. The independent variable is x, the longitudinal coordinate along the string, and the equation of 'motion' is

$$\frac{d^2y}{dx^2} = u; \quad y(0) = y(N) = 0,$$

i.e.

$$A = \begin{bmatrix} 0 & 1 \\ 0 & 0 \end{bmatrix}, \; b = \begin{bmatrix} 0 \\ 1 \end{bmatrix}.$$

If we would like to induce relatively large deflections along the inter= mediate string stations 1,2,...,N-1, with a modest total applied load input (i.e. $\int_0^N u^2(x)dx$ small) then a suitable performance criterion to be minimized is

$$\tilde{J}(u) = \int_0^N u^2(x)dx + \sum_{i=1}^N z'(i)Qz(i),$$

where

$$z \triangleq \binom{y}{y'}$$

and

$$Q = \begin{bmatrix} -1 & 0 \\ 0 & 0 \end{bmatrix}.$$

A similar analysis could have been carried out for an elastic beam struc= ture. Here, we are after the optimal load distribution $u^*(x)$ and the corresponding deflections $y^*(i)$, i=1,2,...,N; evidently a feedback solu= tion is of little interest.

Thus, in this section we derive the open-loop optimal control and corres= ponding trajectory in the hybrid optimal control problem (11.2.1), (11.2.2). Indeed, employing a 'finite element' approach (in the terminology of our previous example) we obtained the discrete optimal control problem. (11.2.4), (11.2.5). Exploiting the fact that if the pair (A,B) is con= trollable then the control matrix $e^{AT}W$ is non-singular, we can now sub= stitute ξ_i from (11.2.5) into (11.2.4) and obtain the minimization pro= blem

$$\min_{x_1,x_2,\ldots,x_N} \sum_{i=1}^{N} x_i'(TQ+e^{-A'T}W^{-1}e^{-AT})x_i+x_{i-1}'W^{-1}x_{i-1}-x_i'e^{-A'T}W^{-1}x_{i-1}$$
$$-x_{i-1}'W^{-1}e^{-AT}x_i. \qquad (11.7.1)$$

Differentiation yields

$$(TQ+e^{-A'T}W^{-1}e^{-AT}+W^{-1})x_i-e^{-A'T}W^{-1}x_{i-1}-W^{-1}e^{-AT}x_{i+1} = 0,\ i=1,2,\ldots,N-1.$$

$$(TQ+e^{-A'T}W^{-1}e^{-AT})x_N-e^{-A'T}W^{-1}x_{N-1} = 0 \qquad (11.7.2)$$

and x_o is given .

Set

$$P \triangleq W^{-1}$$
$$R \triangleq TQ + e^{-A'T}W^{-1}e^{-AT}$$
$$S \triangleq W^{-1}e^{-AT}.$$

The system (11.7.2) can then be written as

$$\bar{A}\bar{x} = \bar{b}, \qquad (11.7.3)$$

where

$$\bar{x} = \begin{bmatrix} x_1 \\ x_2 \\ \vdots \\ \vdots \\ x_N \end{bmatrix}_{Nn\times 1}, \quad \bar{b} = \begin{bmatrix} S' \\ 0 \\ \vdots \\ \vdots \\ 0 \end{bmatrix} x_o \quad Nn\times n$$

and $\bar{A}$ is the block-tridiagonal matrix

$$\bar{A} \triangleq \begin{bmatrix} P+R & -S & & & \\ -S' & P+R & -S & & \\ & & \ddots & & \\ & & -S' & P+R & -S \\ & & & -S' & R \end{bmatrix}_{Nn\times Nn}$$

We thus have the following theorem.

THEOREM 11.7.1 *Given $x_0 \in \mathbb{R}^n$, a necessary and sufficient condition for the existence of an optimal solution to the hybrid optimal control problem* (11.2.1), (11.2.2) *is that $\bar{A}$ is non-negative definite and $\bar{b} \in \text{Range}(\bar{A})$; a sufficient condition for the existence of an optimal solution for all $x_0 \in \mathbb{R}^n$, is that $\bar{A}$ is a positive-definite matrix. Also, there exists a unique optimal solution, for all $x_0 \in \mathbb{R}^n$, iff $\bar{A}$ is a positive-definite matrix.*

Verification of the positive-definiteness of $\bar{A}$ (and hence of the existence of a unique optimal solution to the hybrid optimal control problem) is facilitated by the block-tridiagonal structure of the system matrix $\bar{A}$ and Theorem 1 from [9], which we apply recursively. Specifically, we have the following.

THEOREM 11.7.2 [8] *A necessary and sufficient condition for the block-tridiagonal matrix $\bar{A}$ to be positive-definite is that all elements of the matrix sequence M_i, i=1,2,...,N, should be positive-definite; the matrix sequence M_i is given recursively as a solution to the Riccati matrix difference equations*

$$\begin{cases} M_1 = P + R \\ M_i = P + R - S'M_{i-1}^{-1}S, \quad i=2,3,\dots,N-1 \\ M_N = R - S'M_{N-1}^{-1}S \end{cases} \tag{11.7.4}$$

or

$$\begin{cases} M_1 & = R \\ M_i & = P + R - SM_{i-1}^{-1}S', \quad i=2,3,\dots,N. \end{cases} \tag{11.7.5}$$

□

Note that the recursive equations (11.7.4) or (11.7.5) are simpler than the Riccati equations (11.3.4) or (11.3.6); in fact, the number of operations required to ascertain existence of a solution via (11.7.4) or (11.7.5) is 40% less than the number of operations required to ascertain existence via the Riccati recursion (11.3.4).

The solution $(M_i)_{i=1}^{N}$ to the (existence/uniqueness) recursions (11.7.4) or (11.7.5) also proves useful for computing the optimal control. In this respect, the following holds.

THEOREM 11.7.3 [8] *Let*

$$L_i \triangleq -S'M_{i-1}^{-1}, \tag{11.7.6}$$

where

$$M_i, \quad i=2,3,\dots,N,$$

is the solution to the recursion (11.7.5). *There exists a lower block-diagonal-upper block-diagonal decomposition (see also* [10] *and* [11],p.196) *of the block-tridiagonal matrix* $\bar{A}$,

$$\bar{A} = \bar{L}\bar{U},$$

where

$$\bar{L} \triangleq \begin{bmatrix} I & 0 & \dots & & & 0 \\ L_2 & I & 0 & \dots & & . \\ 0 & L_3 & \dots & & & . \\ & & & & & \\ & & & & I & 0 \\ 0 & 0 & & & L_N & I \end{bmatrix} \quad \text{and } \bar{U} \triangleq \begin{bmatrix} M_1 & -S & 0 & \dots & & 0 \\ 0 & M_2 & -S & 0 & \dots & . \\ . & & & & & \\ . & & & & & \\ . & & & & & \\ . & & & & & 0 \\ & & & & & -S \\ 0 & & & & 0 & M_N \end{bmatrix}$$

Note that in view of the lower block-diagonal - upper block-diagonal decomposition of $\bar{A}$ we are thus able to ascertain existence/uniqueness and simultaneously solve the system

$$\bar{A}\bar{x} = \bar{b}$$

for the optimal control. In addition, it can be shown that these recur= sion schemes (11.7.4) - (11.7.6) are numerically stable. We refer the reader to [8] . We remark finally (see [8]) that the computational effort upon solving the recursions (11.7.1) - (11.7.6) can be signifi= cantly reduced in view of a certain adaptation of Froelich's method for tridiagonal systems (see [12],[13] and also eqn (12) in [14]). In fact, for the Riccati approach (11.3.7) (or (11.3.7')), the LU decomposition method (11.7.6) and Froelich's method, the operational counts (the number of multiplications required in order to solve for the trajectory $x_1,\dots,x_N$) are $2.5N[n^3+O(n^2)]$, $2(N-1)[n^3+O(n^2)]$ and $2(N+1)[n^3+O(n^2)]$ respectively. We have assumed here that n^3 multiplications are required for multiplying together two $n\times n$ matrices, as well as for the inversion of an $n\times n$ matrix; however, if Froehlich's recursions are used, matrix inversions are avoided.

11.8 REFERENCES

[1] Dorato, P., Levis, A.H., Optimal linear regulators: the discrete time case. *IEEE Trans. Auto.Control*, Vol AC-16, pp 613-620.

[2] Kalman, R.E., Ho, Y.C., Narendra, K.S., Controllability of linear dynamical systems. in *Contributions to Differential Equations* . Vol I, New York, Interscience 1963.

[3] Kleinman ,D.L., Rao, P.K., Continuous-discrete gain transformation methods for linear feedback control. *Automatica*, Vol 13, 1977, pp 425-428.

[4] Kalman, R.E., Contributions to the theory of optimal control. *Bol.Soc.Mat.Mex.*, Vol 5, pp 102-109, 1960.

[5] Dreyfus, S.E., Kan, Y.C., A general dynamic programming solution of discrete-time linear optimal control problems. *IEEE Trans. Automatic Control*, Vol AC-18, 1973, pp 286-289.

[6] Apostol, T.M., *Mathematical Analysis*, p.147. Addison Wesley, Reading, Massachusetts, 1965.

[7] Pachter, M., Jacobson, D.H., Taylor Formulae for the optimal feed= back and cost in parameter-dependent linear-quadratic regulators. *NRIMS Note* No 53, CSIR, Pretoria, October 1978.

[8] Pachter, M., Jacobson, D.H., A discretization of the cost functional in linear-quadratic optimal control. *International Journal of Control*, Vol 30, No 2, 1979, pp 339-361.

[9] Haynsworth, E.V., Determination of the inertia of a partitioned Hermitian matrix. *Linear Algebra and its applications*, Vol 1, 1968, pp 73-81.

[10] Varah, J.M., On the solution of block-tridiagonal systems arising from certain finite-difference equations. *Math of Computation* , Vol 26, 1972, pp 859-868.

[11] Varga, R., *Matrix Iterative Analysis*. Prentice Hall, 1962.

[12] Froehlich, R., On stable methods of matrix factorization to block-tridiagonal matrices, *Rep* GA-7164(I), 1966, *General Atomic*, San Diego, California.

[13] Froehlich, R., On stable methods of matrix factorization for block-tridiagonal matrices. *Rep* GA-7164(II), 1966, *General Atomic*, San Diego, California.

[14] Dorr, F.W., The direct solution of the discrete Poisson equation on a rectangle. *SIAM Review*, Vol 12, 1970, pp 248-263.

12. CONDITIONAL DEFINITENESS OF QUADRATIC FUNCTIONALS

Often one is concerned not with minimizing a given quadratic form or functional, but with ascertaining whether or not the form is positive definite, perhaps subject to some constraints. Such questions arise principally in connection with second-order optimality conditions in mathematical programming and the calculus of variations, and lead, in the latter case, to the well-known conditions of Legendre and Jacobi. This lecture is devoted to the question of the positive definiteness of quadratic forms, in both finite and infinite dimensions, *subject to linear homogeneous inequality constraints*; or more generally, the positive definiteness of quadratic forms on convex cones, and this is what we mean by the general term *conditional definiteness*.

Two aspects of conditional definiteness are treated below. There is a theoretical or structural aspect, concerned with the possibility of a certain decomposition of conditionally definite forms, and then there is the question of providing useful tests for conditional definiteness.

In Section 12.1 we consider conditional positive definiteness and conditional positive semidefiniteness of quadratic forms in $\mathbb{R}^n$, enlarging the scope in Section 12.2 to Legendre forms in Hilbert space, with an application to the determination of constrained focal times of linear-quadratic control systems.

Section 12.4 is really in the nature of an appendix, and can be read independently of the other sections. It collects together standard material, in control formulation, on the Legendre and Jacobi conditions in the calculus of variations of unconstrained quadratic functionals.

12.1 CONDITIONAL DEFINITENESS IN FINITE DIMENSIONS

12.1.1 Copositive matrices

All mathematicians know what a positive definite matrix is, and how a given real symmetric matrix may be tested for positive definiteness or semidefiniteness. A wider class of symmetric matrices, first discus= sed by T.S. Motzkin [1], is the class of copositive matrices, defined below.

If x is a vector in $\mathbb{R}^n$, we shall write $x \geqslant 0$ to mean that every com= ponent of x is non-negative - the set $\{x \in \mathbb{R}^n \mid x \geqslant 0\}$ is called the *positive orthant* of $\mathbb{R}^n$, denoted by R^n_+.

DEFINITION 12.1.1 *A real symmetric* n×n *matrix* C *is said to be coposi= tive if*

$$x'Cx \geqslant 0 \quad \textit{whenever } x \geqslant 0.$$

C *is strictly copositive if*

$$x'Cx > 0 \textit{ whenever } x \geqslant 0,\ x \neq 0.$$

The following theorems give tests for copositivity - all require that every *principal submatrix* of C be examined in some way.

THEOREM 12.1.2 (Motzkin [2]) *A real symmetric matrix* C *is strictly copositive iff for every principal submatrix* $\bar{C}$ *of* C, *either* $\det\bar{C} > 0$ *or some entry in the last row of* $\bar{C}$ *has a non-positive cofactor in* $\bar{C}$. □

THEOREM 12.1.3 (Gaddum [3]) *A real symmetric matrix* C *is copositive iff for every principal submatrix* $\bar{C}$ *of* C *the system of inequalities*

$$\bar{C}z \geqslant 0, \quad z \geqslant 0$$

admits a non-zero solution z *(of appropriate dimension).* □

THEOREM 12.1.4 (Cottle, Habetler, Lemke [4]) *A real symmetric matrix* C *is copositive iff no principal submatrix* $\bar{C}$ *of* C *has an eigenvector* z *with positive components corresponding to a simple negative eigenvalue.* □

THEOREM 12.1.5 (Martin [5]) *A real symmetric matrix* C *is copositive iff for no nonsingular principal submatrix* $\bar{C}$ *of* C *does the unique solution* z *of the linear equations* $\bar{C}z = -e$ *have only positive components (here* e *denotes the vector (of appropriate dimension) with unit components).* □

There are two easily recognizable special classes of copositive matrices. Every positive semidefinite (ps) matrix S is obviously copositive, while equally obviously, a matrix N having only *non-negative entries* is also copositive. Consequently any matrix C which can be decomposed in the form

$$C = S + N \quad , \text{C ps and N non-negative,} \qquad (12.1.1)$$

must be copositive. The converse question was considered by Diananda [6], with the following somewhat surprising result.

THEOREM 12.1.6 *For dimensions* $n \leq 4$, *every copositive matrix* C *can be decomposed in the form* (12.1.1). *However indecomposable copositive matrices exist for* $n \geq 5$. □

A counter-example (see [6]) in five dimensions was constructed by E.Horn.

For 2×2 matrices it is easy to see that every copositive matrix is either non-negative or ps. For suppose

$$C = \begin{bmatrix} a & b \\ b & c \end{bmatrix}$$

is copositive, so that $x \geq 0$, $y \geq 0$ imply

$$ax^2 + 2bxy + cy^2 \geq 0.$$

Choosing $(x,y) = (1,0)$ and $(0,1)$ we see that the diagonal entries a and c must be non-negative. Next choosing $x = \sqrt{c}$ and $y = \sqrt{a}$ we have

$$0 \leq ac + 2b\sqrt{a}\sqrt{c} + ac = 2\sqrt{ac}\,(\sqrt{ac} + b), \text{ whence } b \geq -\sqrt{ac}\ .$$

It follows that either $b > +\sqrt{ac}$, in which case C is non-negative, or that $b^2 \leq ac$, in which case C is ps.

12.1.2 A-conditional definiteness

A natural generalization of the notion of copositivity arises when we replace the condition $x \geq 0$ by a more general set of linear homogeneous inequalities.

DEFINITION 12.1.7 Let A *be a given real* m×n *matrix. A real symmetric* n×n *matrix* Q *will be said to be A-conditionally positive semidefinite* (A-cps) *if*

$$x'Qx \geq 0 \textit{ whenever } Ax \geq 0,$$

and A-conditionally positive definite (A-cpd) *if*

$$x'Qx > 0 \textit{ whenever } Ax \geq 0,\ x \neq 0.$$

Once again, as was first noted by Jacobson [7], there are two immedia= tely identifiable subcones of A-cps matrices.

First, any ps matrix S is ipso facto A-cps. Second, a matrix of the form of a product A'CA with C being an m×m *copositive* matrix, is A-cps, for then

$$Ax \geq 0 \text{ implies } x'(A'CA)x = (Ax)'C(Ax) \geq 0.$$

Consequently, any matrix Q which can be decomposed in the form

$$Q = A'CA + S, \quad C \text{ copositive }, \; S \text{ ps} \tag{12.1.2}$$

is A-cps. For conditional positive definiteness, a corresponding suf= ficient condition is that Q can be decomposed in the form

$$Q = A'CA + S \tag{12.1.3}$$

with C strictly copositive and S ps and pd on ker A[1].

Once again the converse question arises as to the necessity or other= wise of the existence of such decompositions, and for the case of A-cpd matrices, there is an unqualified affirmative answer.

THEOREM 12.1.8 (Martin, Jacobson [8]). *For any matrix* A, *every* A-cpd *matrix* Q *admits a decomposition* (12.1.3).

The proof of this result uses a modification of a theorem of P.Finsler on pairs of quadratic forms, and is sketched in a more general setting in Section 12.2.

For A-cps matrices, the situation is not so simple, as the following example shows.

EXAMPLE 12.1.9 With $m = n = 2$, consider the matrices

$$A = \begin{bmatrix} 1 & 0 \\ -1 & 0 \end{bmatrix}, \qquad Q = \begin{bmatrix} 0 & 1 \\ 1 & 0 \end{bmatrix}.$$

Then $Ax \geq 0 \Rightarrow x_1 = 0 \Rightarrow x'Qx = 0$, so that Q is A-cps. However for any

---

1) $x'Sx \geq 0$, and $Ax = 0,\ x \neq 0 \Rightarrow x'Sx > 0$.

matrix C, the matrix Q - A'CA is of the form

$$\begin{bmatrix} * & 1 \\ 1 & 0 \end{bmatrix},$$

which cannot be ps. Thus Q admits no decomposition of the form (12.1.2).

In [8] it is shown that each of the following three extra conditions on the matrix A ensures that A-cps matrices must admit a decomposition (12.1.2):

(a) rank A = m;

(b) n=2 and Slater's condition holds;[2)]

(c) $m \leqslant 4$ and Slater's condition holds.

Since (a) implies Slater's condition, one might suspect that Slater's condition would suffice to ensure decomposability of all A-cps matrices. To this conjecture, however, a counterexample having n=3 and m=5, is given in [8].

A somewhat different approach, initiated by M.J.D. Powell, is taken in the forthcoming paper [9]. Although the set of all m×m copositive matrices is a closed convex cone, it does not follow that the cone

$$C_A \triangleq \{M = A'CA \mid C \text{ copositive}\}$$

is necessarily closed, and this turns out to be of importance.

2) Slater's condition is well-known in the mathematical programming literature, and requires that there exist $x \in \mathbb{R}^n$ such that $Ax > 0$ (componentwise).

THEOREM 12.1.10 [9] *A necessary and sufficient condition on a matrix* A *that every* A-cps *matrix be decomposable in the form* (12.1.2) *is that first, the cone* C_A *be closed, and second, either Slater's condition holds or* rank A = n. □

It follows ofcourse that each of the conditions (a), (b),(c) above guarantees that C_A is closed.

Finally we mention that direct tests for conditional definiteness, un= related to the decomposition question, are given in [5].

12.2 CONDITIONAL DEFINITENESS IN HILBERT SPACE

12.2.1 Formulation

With a view to possible application to problems of optimal control, in [10] the problem of conditional definiteness has been formulated and studied in infinite dimensions.

Let X and Y denote two real Hilbert spaces, with a bounded linear operator

$$A : X \to Y,$$

and a closed convex cone $\Gamma \subseteq Y$ being given.

DEFINITION 12.2.1 *A quadratic form* $Q(\cdot)$ *on* X *will be said to be conditionally positive definite* (cpd) *relative to the pair* (A,Γ) *if*

$$Q(x) > 0 \text{ whenever } Ax \in \Gamma,\ x \neq 0.$$

DEFINITION 12.2.2 *A quadratic form* $Q(\cdot)$ *on* X *will be said to be* (A,Γ) *- decomposable if there are quadratic forms* $C(\cdot)$ *on* Y *and* $S(\cdot)$ *on* X *such that*

(i) $C(y) > 0$ *whenever* $y \in \Gamma$, $y \neq 0$;

(ii) $S(\cdot)$ *is* ps on X, *and* $S(x) > 0$ *for* $x \in \ker A$, $x \neq 0$;

(iii) $Q(x) \equiv C(Ax) + S(x)$.

Note that setting $X = \mathbb{R}^n$, $Y = \mathbb{R}^m$, $\Gamma = \mathbb{R}^m_+$ specializes these concepts to those treated in Section 12.1.2, and that as previously, decomposa= bility is an immediate sufficient condition for conditional definite= ness.

THEOREM 12.2.3 *If a quadratic form* Q *on* X *is* (A,Γ)- *decomposable, then* Q *is* cpd *relative to* (A,Γ). □

12.2.2 Legendre forms

A converse to Theorem 12.2.3 was proved by Martin [10] for the class of quadratic forms known variously as *Legendre, elliptic* or *coercive forms*, the central role of which in the calculus of variations was first emphasized by M.R. Hestenes [11].

DEFINITION 12.2.4 *A quadratic form* $Q(\cdot)$ *on a Hilbert space* X *is said to be strongly positive definite if there exists* $\gamma > 0$ *such that*

$$Q(x) \geqslant \gamma \|x\|^2 \text{ for all } x \in X.$$

$Q(\cdot)$ *is said to be completely continuous if whenever a sequence* $x_n \in X$ *converges weakly to an element* x, *written* $x_n \rightharpoonup x$, *then* $Q(x_n) \to Q(x)$.

DEFINITION 12.2.5 *A Legendre form* $Q(\cdot)$ *on a Hilbert space* X *is a quadratic form which can be written as a sum* $Q = Q_+ + Q_0$, *with* Q_+ *strongly* pd *and* Q_0 *completely continuous.*

In finite dimensions all quadratic forms are Legendre forms, and in general, Legendre forms possess many nice properties normally associated with the finite-dimensional case. Among these properties, we mention that if a Legendre form Q is cpd relative to (A,Γ), then Q is *strongly* cpd relative to (A,Γ), i.e. there will exist $\gamma > 0$ such that

$$Q(x) \geqslant \gamma\|x\|^2 \text{ whenever } Ax \in \Gamma.$$

The reader is referred to [11] for a fascinating treatment of Legendre forms. □

12.2.3 Necessity theorems

Using the same basic strategy of proof as was used in [8], the follow= ing results were derived in [10].

THEOREM 12.2.6 *If a Legendre form* $Q(\cdot)$ *on a Hilbert space* X *is cpd relative to a pair* (A,Γ), *then* Q *must be* (A,Γ)*-decomposable. Further-more, the forms* $C(\cdot)$ *on* Y *and* $S(\cdot)$ *on* X *in the decomposition may be taken to be Legendre forms.*

THEOREM 12.2.7 *A necessary and sufficient condition for a Legendre form* $Q(\cdot)$ *on* X *to be cpd relative to a pair* (A,Γ) *is that for some* $\nu > 0$

(a) *the form* $Q(x) + \nu\|Ax\|^2_Y$ *is strongly pd on* X, *and*

(b) *the form*

$$C_\nu(y) \triangleq \min_{x \in X} [Q(x) + \nu\|y-Ax\|^2_Y]$$

satisfies

$$C_\nu(y) > 0 \text{ whenever } y \in \Gamma,\ y \neq 0.$$

Note that Theorem 12.1.8 is simply the finite-dimensional case in Theo= rem 12.2.6. While Theorem 12.2.6 settles the decomposability question,

Theorem 12.2.7 turns out to be more useful in applications. The proofs of these theorems are sketched in the next subsection.

12.2.4 Finsler's theorem and proofs of main theorems

In 1937, P. Finsler [12] proved a theorem which has since been rediscovered several times: if Q_1 and Q_2 are two real quadratic forms in n variables which vanish simultaneously only at the origin, then there exist α,β such that $\alpha Q_1 + \beta Q_2$ is positive definite. Hestenes [13] contains one extension of this result to infinite dimensions, and the proof of Theorem 12.2.7 depends on the following different extension.

THEOREM 12.2.8 [10] *Let* Q_1 *and* Q_2 *be quadratic forms on a Hilbert space* H, *such that* Q_2 *is ps and* $Q_1 + Q_2$ *is a Legendre form, and let* $C \subseteq H$ *be a closed convex cone. Then*

$$Q_1(h) > 0 \quad \textit{whenever } h \in C\setminus\{0\} \quad \textit{and} \quad Q_2(h) = 0 \qquad (12.2.1)$$

iff there exists $\nu > 0$ *such that*

$$Q_1(h) + \nu\, Q_2(h) > 0 \quad \textit{whenever } h \in C \setminus \{0\}. \qquad (12.2.2)$$

PROOF The sufficiency of (12.2.2) for (12.2.1) to hold is immediate. For the converse, suppose (12.2.2) fails to hold for any $\nu > 0$. Then we can find a sequence $\{h_n\} \subseteq C$ with $\|h_n\| = 1$ such that

$$Q_1(h_n) + nQ_2(h_n) \leqslant 0, \quad n=1,2,\ldots \qquad (12.2.3)$$

Being a bounded sequence, $\{h_n\}$ has a weakly convergent subsequence $\{h_{n_k}\}$, $h_{n_k} \xrightarrow{k} h_0$, where, since C is closed and convex, and thus weakly closed, we have

$$h_0 \in C. \qquad (12.2.4)$$

From (12.2.3) we have firstly that

$$Q_2(h_{n_k}) \leqslant - Q_1(h_{n_k})/n_k \leqslant \|Q_1\|/n_k \overset{k}{\to} 0,$$

so that, since Q_2 is ps and hence weakly lower semicontinuous, we must have

$$Q_2(h_o) = 0. \tag{12.2.5}$$

However, it also follows from (12.2.3) that $Q_1(h_n) + Q_2(h_n) \leqslant 0$, and hence by the weak lower semicontinuity of Legendre forms that

$$Q_1(h_o) = Q_1(h_o) + Q_2(h_o) \leqslant \underline{\lim}_k \, [Q_1(h_{n_k}) + Q_2(h_{n_k})] \leqslant 0. \tag{12.2.6}$$

But a Legendre form Q also has the property (see [10]) that

$$\|h_k\| = 1,\ h_k \rightharpoonup 0 \Rightarrow \underline{\lim}_k \, Q(h_k) > 0,$$

and hence (12.2.6) implies that

$$h_o \neq 0. \tag{12.2.7}$$

Relations (12.2.4-7) together show that (12.2.1) also fails. □

We remark that it is not difficult to give an example (with $H=L^2$, say) showing that the hypothesis that $Q_1 + Q_2$ be a Legendre form is not redundant.

Theorems 12.2.6 and 12.2.7 may be proved together as follows. If we agree that 'the ν-condition' will denote the requirement that conditions (a) and (b) in Theorem 12.2.7 hold for some ν, we establish the impli= cations

$$Q \text{ is cpd rel.}(A,\Gamma) \Rightarrow \nu\text{-condition} \Rightarrow Q \text{ is } (A,\Gamma)\text{-decomposable}. \tag{12.2.8}$$

Theorem 12.2.3 then completes the cycle, showing the equivalence of the three conditions in (12.2.8). We refer the reader to [10] for the further proofs that the forms C_ν and S_ν are Legendre.

The first implication in (12.2.8) is a consequence of Theorem 12.2.8, as we now demonstrate. Let Q be a cpd Legendre form on X relative to $(A,\Gamma) \subseteq Y$. Let $H = X \times Y$, and define quadratic forms Q_1, Q_2 on H by

$$Q_1(x,y) = Q(x), \quad Q_2(x,y) = \|y-Ax\|_Y^2.$$

It follows easily that since Q is a Legendre form on X, $Q_1 + Q_2$ is a Legendre form on H, furthermore Q_2 is ps. With $C \triangleq \{(x,y) \in H \mid y \in \Gamma\}$, the statement that Q is cpd relative to (A,Γ) is equivalent to

$$Q_1(h) > 0 \text{ whenever } h \in C \setminus \{(0,0)\} \text{ and } Q_2(h) = 0.$$

By Theorem 12.2.8 it follows that there exists $\nu > 0$ such that

$$Q(x) + \nu\|y-Ax\|_Y^2 > 0 \text{ whenever } y \in \Gamma \text{ and } (x,y) \neq (0,0). \quad (12.2.9)$$

Choosing y=0 it follows that the Legendre form $Q(x) + \nu\|Ax\|_Y^2$ is pd, and hence strongly pd, on X. From this it follows that for any fixed $y \in Y$, the inhomogeneous functional

$$Q(x) + \nu\|y-Ax\|_Y^2 \quad , \quad x \in X$$

is bounded below on X and achieves its bound, and hence it follows from (12.2.9) that if $y \in \Gamma \setminus \{0\}$, then

$$\min_{x \in X} \; [Q(x) + \nu\|y-Ax\|_Y^2] > 0.$$

Thus the ν-condition holds for this value of ν.

Finally we deal with the second implication in (12.2.8). Suppose then that for some $\nu > 0$, conditions (a) and (b) hold. From (b), for any

$x \in X$, $y \in Y$, we have

$$Q(x) + \nu\|y-Ax\|_y^2 \geq C_\nu(y),$$

and, in particular, choosing $y = Ax$, we have

$$Q(x) \geq C_\nu(Ax) \text{ for all } x \in X.$$

It follows that the form

$$S_\nu(x) \triangleq Q(x) - C_\nu(Ax)$$

is ps on X. On ker A(i.e. when Ax=0), $S_\nu(\cdot)$ coincides with $Q(\cdot)$, and hence by (a) we have

$$Ax = 0 \Rightarrow S_\nu(x) = Q(x) + \nu\|Ax\|^2 > 0 \text{ unless } x = 0.$$

Thus, S_ν is pd on ker A, which shows that Q is (A,Γ) -decomposable.

12.3 AN APPLICATION TO CONSTRAINED FOCAL TIMES

12.3.1 The strengthened Legendre condition

We consider a linear control system

$$\dot{x} = A(t)x + B(t)u \quad t_0 \leq t \leq T$$

with homogeneous initial condition

$$x(t_0) = 0,$$

having homogeneous linear inequality constraints upon the final state, i.e. for some r×n matrix D we require

$$Dx(T) \geq 0. \tag{12.3.1}$$

Let

$$J(t_0,u(\cdot)) = \int_{t_0}^{T} (u'(t)R(t)u(t) + x'(t)Q(t)x(t))dt + x'(T)Q_f x(T)$$

be a quadratic functional. We regard the final time T as fixed, and suppose for simplicity that the matrix functions $A(\cdot)$, $B(\cdot)$, $R(\cdot)$, $Q(\cdot)$ are continuous. For given $t_0 < T$, we take as admissible controllers the Hilbert space

$$X = L^2[[t_0,T], \mathbb{R}^m].$$

THEOREM 12.3.1 *A necessary and sufficient condition for* $J(t_0, \cdot)$ *to be a Legendre form on* X *is that* $R(\cdot)$ *be uniformly positive definite on* $[t_0,T]$, *i.e. there exists* $\gamma > 0$ *such that for all* $t \in [t_0,T]$

$$u'R(t)u \geqslant \gamma|u|^2 \textit{ for all } u \in \mathbb{R}^m. \tag{12.3.2}$$

PROOF Owing to the integration involved in passing from a controller to its response, the state terms in J are completely continuous, whereas the control term is not. It follows that J is Legendre iff the latter term is strongly positive definite. Using 'pulse' controllers of the form

$$u(t) = \begin{cases} \frac{1}{\sqrt{\varepsilon}}\bar{u} & \text{if} \quad \tau \leqslant t \leqslant \tau + \varepsilon \\ 0 \text{ otherwise} \end{cases}$$

it follows easily that this is equivalent to (12.3.2). □

REMARK When $R(\cdot)$ is continuous, as we have assumed for simplicity, (12.3.2), is equivalent to the requirement that R(t) be positive definite for all t. However Theorem 12.3.2 remains true even where $R(\cdot)$ is merely bounded and measurable on $[t_0,T]$, ('all t' being replaced by 'almost all t').

J is said to satisfy the *strengthened Legendre condition* when (12.3.2) is satisfied. Its general significance is further elaborated in Section 12.4.

12.3.2 The constrained focal time

We assume henceforth that the strengthened Legendre condition is satis= fied, but that $Q(\cdot)$ and/or Q_f is not ps, so that J is not a priori non-negative, and pose the problem of finding the *focal time* t^*, defined as the infinum of initial times t_0 for which $J(t_0,\cdot)$ is positive defi= nite subject to the inequality constraints (12.3.1) on the final state.

Let $A : X = L^2[[t_0,T], \mathbb{R}^m] \to Y = \mathbb{R}^r$ be the bounded linear operator which assigns to each controller $u(\cdot)$ the vector $Dx(T) \in \mathbb{R}^r$, so that (12.3.1) can be written as

$$Au \in \Gamma \triangleq \mathbb{R}^r_+ .$$

Since J is a Legendre form on X, we may apply our fundamental Theorem 12.2.7 to conclude that for given $t_0 < T$, $J(t_0,\cdot)$ is cpd relative to (12.3.1) iff there exists $\nu > 0$ such that

$$J(t_0,u(\cdot)) + \nu|Dx(T)|^2 \text{ is strongly pd on } X. \quad (12.3.3)$$

while the form on $Y = \mathbb{R}^r$ given by

$$C_\nu(y) = \min_{u(\cdot) \in X} [J(t_0,u(\cdot)) + \nu|y - Dx(T)|^2] \quad (12.3.4)$$

is strongly copositive.

For given ν, the form (12.3.3) is

$$\int_{t_0}^{T} [u'(t)R(t)u(t) + x'(t)Q(t)x(t)]dt + x'(T)(Q_f+\nu D'D)x(T),$$

and it is a standard result (see Theorems 12.4.8 and 12.4.9) below) that this form is strongly pd iff the Riccati problem

$$-\dot{P} = A'(t)P + PA(t) - PB(t)R^{-1}(t)B'(t)P + Q(t) \quad t_0 \leqslant t \leqslant T \tag{12.3.5}$$

$$P(T) = Q_f + \nu D'D$$

has a solution on $[t_0, T]$, i.e. does not 'blow up'.

To evaluate $C_\nu(y)$, it is convenient to introduce a further state vector $z \in \mathbb{R}^r$, satisfying the equation

$$\dot{z} = 0,$$

so that we may regard y as both initial and final state of z. Then (12.3.4) becomes

$$C_\nu(y) = \min_{u(\cdot) \in X} \left\{ \int_{t_0}^{T} [u'(t)R(t)u(t) + x'(t)Q(t)x(t)]\,dt \right.$$

$$\left. + [x'(T), z'(T)] \left[\begin{array}{c|c} Q_f + \nu D'D & -\nu D' \\ \hline -\nu D & \nu I \end{array}\right] \left[\begin{array}{c} x(T) \\ \hline z(T) \end{array}\right] \right\}$$

where $\begin{bmatrix} \dot{x} \\ \dot{z} \end{bmatrix} = \left[\begin{array}{c|c} A(t) & 0 \\ \hline 0 & 0 \end{array}\right] \begin{bmatrix} x \\ z \end{bmatrix} + \begin{bmatrix} B \\ 0 \end{bmatrix} u, \quad \begin{bmatrix} x(t_0) \\ z(t_0) \end{bmatrix} = \begin{bmatrix} 0 \\ y \end{bmatrix}.$

This is a standard unconstrained LQ problem, and hence (see Section 4.3.2 of Lecture 4)

$$C_\nu(y) = [0', y']\tilde{P}(t_0) \begin{bmatrix} 0 \\ y \end{bmatrix} = y'\tilde{P}_{22}(t_0)y, \tag{12.3.6}$$

where

$$\tilde{P}(\cdot) = \left[\begin{array}{c|c} \tilde{P}_{11}(\cdot) & \tilde{P}_{12}(\cdot) \\ \hline \tilde{P}_{21}(\cdot) & \tilde{P}_{22}(\cdot) \end{array}\right]$$

is the solution of the associated Riccati problem. Owing to the presence of blocks of zeros,this Riccati problem decomposes into a non=linear system identical to (12.3.5), to be satisfied by the block $\tilde{P}_{11}(\cdot)$, followed by the linear problems

$$-\dot{\tilde{P}}_{12} = [A(t) - B(t)R^{-1}(t)B'(t)\tilde{P}_{11}(t)]'\tilde{P}_{12} \quad , \tilde{P}_{12}(T) = -\nu D' \qquad (12.3.7)$$

and

$$-\dot{\tilde{P}}_{22} = -\tilde{P}_{12}'(t)B(t)R^{-1}(t)B'(t)\tilde{P}_{12}(t), \quad \tilde{P}_{22}(T) = \nu I. \qquad (12.3.8)$$

For any choice of ν we may integrate equation (12.3.5) (12.3.7-8), and as we do so, test $\tilde{P}_{22}(t)$ for strict copositivity using Motzkin's test (Theorem 12.1.2). In this way we determine the time $t_o(\nu)$ at which either $\tilde{P}_{22}(t)$ first (in the sense of backward time!) loses strict copositivity or $\tilde{P}_{11}(t)$ blows up. The sufficiency part of Theorem 12.2.7 then ensures that the desired focal time t* satisfies

$$t^* \leqslant t_o(\nu).$$

On the other hand, the necessity part ensures that for any $t_o > t^*$, there exists ν such that $t^* \leqslant t_o(\nu) < t_o$, and consequently

$$t^* = \liminf_{\nu \to \infty} t_o(\nu).$$

Finally, we observe that $t_o(\nu)$ is obviously non-increasing, so that in fact

$$t^* = \lim_{\nu \to \infty} t_o(\nu).$$

12.3.3 Example

For the double integrator control system

$$\dot{x}_1 = x_2, \quad \dot{x}_2 = u, \quad x_1(t_o) = x_2(t_o) = 0$$

with final time constraints

$$x_1(0) \geqslant 0 \quad x_2(0) \leqslant 0,$$

we determine the infimum of times $t_0 < 0$ for which

$$\int_{t_0}^{0} (u^2 - x_1^2)dt$$

is cpd. By carrying out the above procedure analytically (see [10] for more details), it is found that the focal time t^* is the first negative zero of the expression

$$\cos t \sinh t + \sin t \cosh t,$$

which is

$$t^* = -2.365020..... \qquad (12.3.9)$$

As a numerical test, equations (12.3.5,7,8) were integrated using a standard Kutta-Merson fourth-order method, checking the strict coposi= tivity of $\tilde{P}_{22}(t)$ every 0,02 time units, and using cubic interpolation to find the time $t_0(\nu)$ at which strict copositivity was lost. With truncation error controlled by a mixed test of the form

$$|\delta\tilde{P}| < 10^{-5}(1 + |\tilde{P}|)$$

for each entry of $\tilde{P}$, the results for $\nu = 1,2,4,8,16,32$ are given in the following table.

ν	$t_o(\nu)$
1	-2.144572
2	-2.225842
4	-2.285386
8	-2.322211
16	-2.342798
32	-2.353695

Finally, using cubic polynomials in ν^{-1}, successive fours of these values were used to extrapolate to the limit as $\nu \to \infty$. These extra= polation results are given in the final table, and are satisfactorily close to the correct value as given by (12.3.9).

Extrapolation from	t_o^*
ν = 1,2,4,8	-2.364624
ν = 2,4,8,16	-2.364991
ν = 4,8,16,32	-2.365018

12.3.4 Closing remarks

A very interesting problem, not touched upon above, and still very much a topic of active research, is the problem of conditional definite= ness subject to conic control constraints. The prototype conditional definiteness problem of this type replaces the final state constraint (12.3.1) by the constraint

$$u(t) \in \mathbb{R}_+^m \text{ for almost all } t \in [t_o, T],$$

and is the direct analogue for LQ control systems of copositivity for matrices. However, with the exception of the special case treated in [14] there are no known useful tests for such conditional definiteness. In [15] a condition involving eigenvalue problems for variational in= equalities is given, but its practical significance is doubtful.

12.4 THE SIMPLEST DEFINITENESS PROBLEM IN LQ CONTROL THEORY

This Section is really an appendix, which can be read independently of the rest of Lecture 12. It presents an account of the Legendre and Jacobi conditions for definiteness of a quadratic functional defined on a linear control system, having zero initial state, free final state, and unconstrained control. This is thus standard nineteenth century second-variation material, presented however in *control* formulation, as first written down by Breakwell and Ho [16].

12.4.1 The Legendre conditions

Consider a quadratic functional

$$J(u(\cdot)) = \int_{t_0}^{T} [u'(t)R(t)u(t) + x'(t)Q(t)x(t)]\,dt + x'(T)Q_f x(T)$$

defined on a control system

$$\dot{x} = A(t)x + B(t)u, \qquad (12.4.1)$$

in which, for simplicity, we suppose the matrix functions $R(\cdot)$, $Q(\cdot)$, $A(\cdot)$ and $B(\cdot)$ to be continuous, and take the homogeneous initial condi= tion

$$x(t_0) = 0, \qquad (12.4.2)$$

so that J is a *homogeneous* quadratic functional of square-integrable controllers. Note that the final state x(T) is free.

DEFINITION 12.4.1 *The functional J is said to be strongly positive definite if there exists* $\gamma > 0$ *such that*

$$J(u(\cdot)) \geqslant \gamma\int_{t_0}^{T} |u(t)|^2 dt \text{ for all } u(\cdot).$$

The terms *positive definite* and *positive semidefinite* have their usual meanings in the sequel.

EXAMPLE 12.4.2 The functional

$$J(u(\cdot)) = \int_0^1 x^2(t)dt \text{ with } \dot{x} = u,\ x(0) = 0,$$

for which $R(t) \equiv 0$, is positive definite, since

$$J(u(\cdot)) = 0 \Rightarrow x(t) \equiv 0 \Rightarrow u(t) \equiv 0.$$

However it follows from Theorem 12.4.3 below that the functional cannot be strongly positive definite.

THEOREM 12.4.3 *For J to be strongly positive definite it is necessary that there exist* $\gamma > 0$ *such that*

$$u'R(t)u \geqslant \gamma|u|^2 \text{ for all } u \in \mathbb{R}^m \text{ and } t \in [t_0,T]. \qquad (12.4.3)$$

For J to be positive semidefinite it is necessary that $R(t)$ *be positive semidefinite for all* $t \in [t_0,T]$.

PROOF For any time $\tau \in [t_0,T)$ and vector $\bar{u} \in \mathbb{R}^m$, held fixed, we con= sider the effect of the 'pulse' controller

$$u(t) = \begin{cases} \frac{1}{\sqrt{\varepsilon}}\,\bar{u} & \text{for } \tau \leqslant t \leqslant \tau + \varepsilon \\ 0 & \text{otherwise in } [t_0,T] \end{cases}$$

for all sufficiently small positive values of ε. First, we have

$$\int_{t_0}^{T} u'(t)R(t)u(t)dt = \frac{1}{\varepsilon} \int_{\tau}^{\tau+\varepsilon} \bar{u}'R(t)\bar{u}dt \ \underset{0}{\overset{\varepsilon}{\rightarrow}} \ \bar{u}'R(\tau)\bar{u}. \qquad (12.4.4)$$

On the other hand, because of (12.4.2), the response $x(\cdot)$ to this pulse controller is

$$x(t) = \begin{cases} 0 & \text{for } t \leqslant \tau \\ \frac{1}{\sqrt{\varepsilon}} \int_{\tau}^{t} \Phi(t,\sigma)B(\sigma)\bar{u}d\sigma & \text{for } \tau \leqslant t \leqslant \tau + \varepsilon \\ \frac{1}{\sqrt{\varepsilon}} \int_{\tau}^{\tau+\varepsilon} \Phi(t,\sigma)B(\sigma)\bar{u}d\sigma & \text{for } \tau + \varepsilon \leqslant t \ . \end{cases}$$

Hence if K denotes an upper bound for $|\Phi(t,\sigma)B(\sigma)\bar{u}|$ on the compact set $t_0 \leqslant t, \sigma \leqslant T$, we have

$$|x(t)| \leqslant \frac{1}{\sqrt{\varepsilon}} (K\varepsilon) = K\sqrt{\varepsilon} \text{ for all } t \in [t_0,T].$$

It follows that

$$\int_{t_0}^{T} x'(t)Q(t)x(t)dt + x'(T)Q_f x(T) \ \underset{0}{\overset{\varepsilon}{\rightarrow}} \ 0,$$

and hence by (12.4.4) that

$$J(u(\cdot)) \ \underset{0}{\overset{\varepsilon}{\rightarrow}} \ \bar{u}'R(\tau)\bar{u}. \qquad (12.4.5)$$

Suppose that J is strongly positive definite. Then there exists $\gamma > 0$ such that

$$J(u(\cdot)) \geqslant \gamma \int_{t_0}^{T} |u(t)|^2 dt = \gamma \frac{1}{\varepsilon} \int_{\tau}^{\tau+\varepsilon} |\bar{u}|^2 dt = \gamma|\bar{u}|^2,$$

and hence it follows from (12.4.5) that also

$$\bar{u}'R(\tau)\bar{u} \geqslant \gamma|\bar{u}|^2.$$

By continuity this holds also for $\tau = T$, and this proves the first assertion of the theorem. The second assertion also follows immedia= tely from (12.4.5). □

The condition that $R(t)$ be positive semidefinite for all $t \in [t_0,T]$ is known as the *Legendre condition* , while the condition that there exist $\gamma > 0$ for which (12.4.3) holds, is called the *strengthened Legendre condition*, and when this holds $R(t)$ is invertible for each $t \in [t_0,T]$.

12.4.2 The Jacobi conditions

In this subsection we derive the control analogue of the classical Jacobi condition of the calculus of variations. Assuming for the moment that $R(t)$ is invertible for each t, we may consider the homo= geneous *Hamiltonian system* (cf. Section 4.3.4)

$$\begin{bmatrix} \dot{x} \\ \dot{\lambda} \end{bmatrix} = \begin{bmatrix} A(t) & -B(t)R^{-1}(t)B'(t) \\ -Q(t) & -A'(t) \end{bmatrix} \begin{bmatrix} x \\ \lambda \end{bmatrix}, \qquad (12.4.6)$$

and consider *non-trivial* solutions $x(\cdot)$, $\lambda(\cdot)$ which satisfy the final time condition

$$\lambda(T) = Q_f x(T). \qquad (12.4.7)$$

Obviously non-trivial solutions satisfying (12.4.7) must have $x(T) \neq 0$, and hence also $x(t) \neq 0$ for t near T.

DEFINITION 12.4.4 *The (backward) focal time* t^* *of* J *is the maximum of times* $\tau < T$ *for which there is a non-trivial solution* $x(\cdot)$, $\lambda(\cdot)$ *of* (12.1.6) *on* $[\tau,T]$ *satisfying* (12.4.7) *and* $x(\tau) = 0$.

THEOREM 12.4.5 *For* J *to be strongly positive definite it is necessary that* $t^* < t_0$. *(Note that* t^* *is then well-defined since* R^{-1} *exists by Theorem* 12.4.3).

PROOF Suppose J is strongly positive definite but that $t_o \leqslant t^*$. Owing to the continuous dependence of solutions on initial data, there is a non-trivial solution $x^*(\cdot)$, $\lambda^*(\cdot)$ of (12.1.6) on $[t^*,T]$ satisfying (12.4.7) and

$$x^*(t^*) = 0. \tag{12.4.8}$$

Consider the controller

$$u(t) = \begin{cases} 0 & \text{for } t_o \leqslant t < t^* \\ -R^{-1}(t)B'(t)\lambda^*(t) & \text{for } t^* \leqslant t \leqslant T. \end{cases} \tag{12.4.9}$$

Using (12.4.2), (12.4.6) and (12.4.8) it is easily checked that the corresponding response is

$$x(t) = \begin{cases} 0 & \text{for } t_o \leqslant t < t^* \\ x^*(t) & \text{for } t^* \leqslant t \leqslant T. \end{cases} \tag{12.4.10}$$

From (12.4.6) we have for all $t \in [t^*,T]$:

$$\frac{d}{dt}(\lambda^{*\prime}x^*) = (-Qx^* - A'\lambda^*)'x^* + \lambda^{*\prime}(Ax^* - BR^{-1}B'\lambda^*)$$

$$= -x^{*\prime}Qx^* - u'Ru.$$

Using the above relations it follows that

$$J(u(\cdot)) = \int_{t^*}^{T} (u'Ru + x^{*\prime}Qx^*)dt + x^{*\prime}(T)Q_f x^*(T)$$

$$= -\lambda^{*\prime}(T)x^*(T) + \lambda^{*\prime}(t^*)x^*(t^*) + x^{*\prime}(T)Q_f x^*(T)$$

$$= 0 \text{ by (12.4.7) and (12.4.8).} \tag{12.4.11}$$

Since J is strongly positive definite it follows that $u^*(t) \equiv 0$, which implies that the response (12.4.10) vanishes identically. Using (12.4.7) it follows then that also $\lambda^*(T) = 0$, which contradicts the stipulation in the definition of t^* that the solution of (12.4.6) be non-trivial. □

The condition $t^* < t_0$ is called the *strengthened Jacobi condition* - the unstrengthened form requires merely $t^* \leqslant t_0$.

The counterpart of this theorem for positive semidefiniteness involves an additional controllability hypothesis.

THEOREM 12.4.6 *Suppose that* $R(t)$ *is invertible on* $[t_0,T]$ *(so that the Hamiltonian system is defined) but that* $t_0 < t^*$, *and that the system* (12.4.1) *is completely controllable over the interval* $[t_0,t^*]$. *Then* J *is not positive semidefinite.*

PROOF Let $x^*(\cdot)$, $\lambda^*(\cdot)$ be as in the previous proof, and consider the controller (12.4.9) with response (12.4.10). By (12.4.11), if J is positive semidefinite, this controller must be an optimal control, and hence must satisfy the Pontryagin maximum principle. We show under the stated controllability hypothesis that this is not the case. As re= marked in Lecture 6, since the terminal state is free, we may choose $\lambda_0 = 1$ in the statement (Theorem 6.2.1) of the maximum principle. It follows easily from the principle that if $u(\cdot)$, $x(\cdot)$ is optimal, the associated multiplier function $\lambda(t)$, $t \in [t_0,T]$ must, together with $x(\cdot)$, satisfy (12.4.6) and (12.4.7) and $u(t) = -R^{-1}(t)B'(t)\lambda(t)$. It follows that $\lambda(t) \equiv \lambda^*(t)$ on $[t^*,T]$, while on $[t_0,t^*]$ we have $\dot{\lambda}(t) = -A'(t)\lambda(t)$ and $B'(t)\lambda(t) \equiv 0$, so that

$$B'(t)\Phi'(t^*,t)\lambda(t^*) \equiv 0 \text{ on } [t_0,t^*].$$

Because of the complete controllability hypothesis this implies $\lambda(t^*)=0$, i.e. $\lambda^*(t^*) = 0$. Since also $x^*(t^*) = 0$, this contradicts the non-triviality of the solution $x^*(\cdot)$, $\lambda^*(\cdot)$ of (12.4.6). This ends the proof.□

COROLLARY 12.4.7 *If the system* (12.4.1) *is arbitrary-interval controllable and* $R^{-1}(t)$ *exists on* $[t_0,T]$, *then the (unstrengthened) Jacobi condition is necessary for* J *to be positive semidefinite.*

The next theorem characterizes the focal time t^* as the 'blow-up' time of the Riccati differential equation

$$-\dot{P} = A'(t)P + PA(t) - PB(t)R^{-1}(t)B'(t)P + Q(t) \qquad (12.4.11)$$

with final condition

$$P(T) = Q_f. \qquad (12.4.12)$$

If the solution $P(\cdot)$ to this problem exists on an interval $(\hat{t},T]$, but $\|P(t)\|\uparrow\infty$ as $t\downarrow\hat{t}$, we refer to $\hat{t}$ as the blow-up time of the problem (12.4.11-12).

THEOREM 12.4.8 *The focal time* t^* *of the Hamiltonian problem* (12.4.6-7) *coincides with the blow-up time* $\hat{t}$ *of the Riccati problem* (12.4.11-12).

PROOF As in Section 4.3.4, let $X(\cdot)$, $\Lambda(\cdot)$ denote the solution of the problem

$$\begin{bmatrix}\dot{X}\\ \dot{\Lambda}\end{bmatrix} = \begin{bmatrix}A(t) & -B(t)R^{-1}(t)B'(t)\\ -Q(t) & -A'(t)\end{bmatrix}\begin{bmatrix}X\\ \Lambda\end{bmatrix} \qquad t_0\leqslant t\leqslant T$$

with final condition

$$\begin{bmatrix}X(T)\\ \Lambda(T)\end{bmatrix} = \begin{bmatrix}I_n\\ \hline Q_f\end{bmatrix}.$$

Then the general non-trivial solution of (12.4.6-7) is of the form

$$\begin{bmatrix} x(t) \\ \\ \lambda(t) \end{bmatrix} = \begin{bmatrix} X(t) \\ \\ \Lambda(t) \end{bmatrix} c = \begin{bmatrix} X(t)c \\ \\ \Lambda(t)c \end{bmatrix}$$

where c is an arbitrary non-zero constant vector in $\mathbb{R}^n$. Consequently to have $x(\tau) = 0$ we must have $X(\tau)c = 0$, so that $X(\tau)$ is singular. Conversely, if $X(\tau)$ is singular, we may choose any $c \neq 0$ in $\ker X(\tau)$ to produce a solution of (12.4.6-7) having $x(\tau) = 0$. It follows that the focal point t^* is precisely that time at which $X(\cdot)$ first (in the backwards sense!) becomes singular. It follows as in Theorem 4.3.3 that

$$P(t) \triangleq \Lambda'(t)X^{-1}(t)$$

is a solution of the Riccati problem (12.4.11-12) defined on $(t^*,T]$, and it remains only to show that $\|P(t)\| \uparrow \infty$ as $t \downarrow t^*$. Let $c \neq 0$ be such that $X(t^*)c = 0$. Then for $t \in (t^*,T]$

$$P(t)X(t)c = \Lambda(t)c,$$

or

$$P(t)x(t) = \lambda(t). \qquad (12.4.13)$$

where $x(\cdot)$, $\lambda(\cdot)$ is a *non-trivial* solution of the Hamiltonian system, such that $x(t^*) = 0$. Hence we cannot also have $\lambda(t^*) = 0$, so that (12.4.13) implies that $\|P(t)\| \uparrow \infty$. □

12.4.3 Sufficiency of the strengthened conditions

THEOREM 12.4.9 *For J to be strongly positive definite it is necessary and sufficient that the strengthened Legendre and strengthened Jacobi conditions hold.*

PROOF The necessity has been proved in Theorem 12.4.3 and 12.4.5 above. Suppose, for the sufficiency, that the strengthened Legendre and Jacobi conditions hold. Then, for all sufficiently small $\varepsilon > 0$, the matrix

$$R_\varepsilon(t) = R(t) - \varepsilon I_m$$

is positive definite on $[t_0,T]$, and the Riccati equation

$$-\dot{P} = A'(t)P + PA(t) - PB(t)R_\varepsilon^{-1}(t)B'(t)P + Q(t)$$

with final condition (12.4.12) will also have a solution $P_\varepsilon(\cdot)$ defined on $[t_0,T]$. Hence if we use this solution just as in Section 4.3.2 of Lecture 4, we may transform the functional

$$J_\varepsilon(u(\cdot)) \triangleq \int_{t_0}^{T} [u'(t)R_\varepsilon(t)u(t) + x'(t)Q(t)x(t)]\,dt + x'(T)Q_f x(T)$$

into the following form (noting that $x_0 = x(t_0) = 0$):

$$J_\varepsilon(u(\cdot)) = \int_{t_0}^{T} (u + R_\varepsilon^{-1}B'P_\varepsilon x)'R_\varepsilon(u + R_\varepsilon^{-1}B'P_\varepsilon x)\,dt \geqslant 0.$$

Hence

$$J(u(\cdot)) = J_\varepsilon(u(\cdot)) + \varepsilon \int_{t_0}^{T} |u(t)|^2 dt \geqslant \varepsilon \int_{t_0}^{T} |u(t)|^2 dt$$

which shows that J is strongly positive definite. □

12.5 REFERENCES

[1] T.S. Motzkin, in *Nat. Bur. Standards Rep.* 1818, pp 11-12, 1952.

[2] T.S. Motzkin, Signs of minors. In *Inequalities* (O.Shisha, Ed.), Academic Press, New York, 1967.

[3] J.W. Gaddum, Linear inequalities and quadratic forms. *Pacific J. Math.*, 8, pp. 411-414, 1958.

[4] R.W. Cottle, G.J. Habetler and C.E. Lemke, On classes of copositive matrices. *Lin. Alg. and its Appl.*, Vol 3, pp 295-310.

[5] D.H. Martin, Finite criteria for conditional definiteness of quad= ratic form. *NRIMS Technical Report* TWISK 115, Pretoria, August 1979.

[6] P.H. Diananda, On non-negative forms in real variables some or all of which are non-negative. *Proc. Cambridge Philos. Soc.*, Vol 58, pp 17-25, 1962.

[7] D.H. Jacobson, *Extensions of Linear-Quadratic Control, Optimization and Matrix Theory*. Academic Press, New York, 1977.

[8] D.H. Martin and D.H. Jacobson, Copositive matrices and definiteness of quadratic forms subject to homogeneous linear inequality con= straints. *NRIMS Technical Report* TWISK 89, Pretoria, May 1979.

[9] D.H. Martin, M.J.D. Powell and D.H. Jacobson, On the decomposition of conditionally positive semidefinite matrices. forthcoming NRIMS Technical Report.

[10] D.H. Martin, Conditional positivity of quadratic forms in Hilbert space. *NRIMS Technical Report* TWISK 113, Pretoria, August 1979.

[11] M.R. Hestenes, Applications of the theory of quadratic forms in Hilbert space to the calculus of variations. *Pacific J. Math.* Vol 1(4), pp 525-581, 1951.

[12] F. Finsler, Über das Vorkommen definiter und semidefiniter Formen in Scharen quadratischer Formen. *Comm. Math. Helvetia*, Vol 9, pp 188-192, 1937.

[13] M.R. Hestenes, *Optimization Theory*. J. Wiley & Sons, New York, 1975.

[14] D.H. Martin and D.H. Jacobson, Optimal control laws for a class of constrained linear-quadratic problems. *Automatica*, Vol 15, pp 431-440, 1979.

[15] D.H. Martin, The spectra of bounded linear self-adjoint operators relative to a cone in Hilbert space. *NRIMS Technical Report* TWISK 30, Pretoria, May 1978.

[16] J.V. Breakwell and Yu-Chi Ho, On the conjugate point condition for the control problem. *Int.J. Engng. Sci.* , Vol 2, pp 565-579, 1965.

13. EXPONENTIAL PERFORMANCE CRITERIA AND STATE-DEPENDENT NOISE

13.1 INTRODUCTION

In this lecture we consider the control of stochastic discrete-time linear dynamic systems with exponential performance criteria, and a certain class of stochastic discrete-time nonlinear dynamic systems with quadratic performance criteria. On the one hand we extend the usual linear-quadratic-stochastic control problem by replacing the quadratic criterion by an exponential, while on the other hand we retain the quadratic criterion and alter the dynamic system.

13.2 LINEAR-QUADRATIC-STOCHASTIC CONTROL PROBLEMS

For completeness we briefly review the standard stochastic linear-quadratic problem [1].

We consider minimizing the expected value of the quadratic function ψ, where

$$2\psi = \sum_{k=o}^{N-1} (x_k' Q_k x_k + u_k' R_k u_k) + x_N' Q_N x_N \qquad (13.2.1)$$

subject to the stochastic discrete-time linear system

$$x_{k+1} = A_k x_k + B_k u_k + \Gamma_k \omega_k \quad , \qquad x_o \text{ given,} \qquad (13.2.2)$$

where $Q_k \geqslant 0$, $k = 0,\ldots,N$, $R_k > 0$, $k=0,\ldots,N-1$, and where the independent (not necessarily Gaussian) sequence $\{\omega_k \in R^q\}$ has the statistics

$$E[\omega_k] = 0, \quad E[\omega_j \omega_k'] = \Lambda_k \, \delta_{jk} \qquad (13.2.3)$$

with $\Lambda_k \geqslant 0$, and where $\delta_{jk} = 1$ if $j=k$ and is zero otherwise.

We assume also that x_k, $k=0,\dots,N$, can be measured exactly.

In this case it turns out that the optimal controller is the same as that for the deterministic system obtained by removing the term $\Gamma_k \omega_k$ in (13.2.2), viz.

$$u_k^* = - \tilde{R}_k^{-1} \tilde{A}_k x_k \quad , \quad k=0,\dots,N-1, \tag{13.2.4}$$

where

$$\tilde{A}_k = B_k' P_{k+1} A_k \quad , \tilde{R}_k = R_k + B_k' P_{k+1} B_k, \tag{13.2.5}$$

$$P_k = Q_k + A_k' P_{k+1} A_k - \tilde{A}_k \tilde{R}_k^{-1} \tilde{A}_k, \quad P_N = Q_N, \tag{13.2.6}$$

and the optimal value of $J = E[\psi]$ is given by

$$\tfrac{1}{2} x_o' P_o x_o + e_o \tag{13.2.7}$$

where

$$e_k = \tfrac{1}{2} tr(P_{k+1} \Gamma_k \Lambda_k \Gamma_k') + e_{k+1}, \quad e_N = 0. \tag{13.2.8}$$

Note that the covariance matrix Λ_k of the system noise does not affect the optimal control law but only the optimal value of J through e_o. Furthermore, as indicated at the beginning of this section, $\{\omega_k\}$ need not be a Guassian sequence, though this is often assumed in the litera= ture.

When x_k cannot be measured directly and only a noisy observation

$$y_k = H_k x_k + \nu_k \tag{13.2.9}$$

is available, the situation is a little more complicated. In this case we assume that

$$E[x_0] = 0, \quad E[x_0 x_0'] = X_0 \tag{13.2.10}$$

and that

$$E[\nu_k] = 0, \quad E[\nu_j \nu_k'] = V_k \delta_{jk} \quad , \quad V_k > 0. \tag{13.2.11}$$

We assume further that the noise sequences $\{\omega_k\}$, $\{\nu_k\}$ and the initial point x_0 are independently distributed and independent of each other, but we do not need to assume that these noises are Gaussian.

Here it turns out that the optimal controller which minimizes the ex= pected value of ψ is given by

$$u_k^* = -\tilde{R}_k^{-1} \tilde{A}_k \hat{x}_k \tag{13.2.12}$$

where x_k (which cannot here be measured) is replaced by

$$\hat{x}_k \triangleq E[x_k | y_k, y_{k-1}, \ldots, y_0]. \tag{13.2.13}$$

Now, if all the noises are assumed to be Gaussian, we find that $\hat{x}_k$ is given by the Kalman (i.e. optimal linear) filter. The linear-quadratic-Gaussian control problem therefore enjoys the so-called separation property, viz. the optimal controller gains are independent of the noise sequences and are pre-computable; the Kalman filter depends only on the noise covariances and not on the optimal controller and is also pre-computable; the optimal controller is then a cascade of the linear optimal filter and the deterministic linear optimal controller.

Analogous results to those described above hold in the continuous-time case [1].

13.3 EXPONENTIAL PERFORMANCE CRITERION

We here consider the consequences of minimizing the expected value of e^{ψ}, rather than the expected value of ψ, in the case of perfect measurement of x_k. At the end of this section we briefly comment upon the results that are obtained when only a noisy measurement of x_k is available. A full treatment of this problem is available in [2].

An obvious fact is that when the term $\Gamma_k\omega_k$ is absent from (13.2.2) - the deterministic case - minimizing ψ is equivalent to minimizing e^{ψ}; however, in the stochastic case $E[\psi]$ and $E[e^{\psi}]$ are not minimized by the same controller. Indeed, $E[e^{\psi}]$ will not exist if the noise covariances $\{\Lambda_k\}$ are too large, while $E[\psi]$ always exists.

First we introduce for convenience the following definitions:

$$X_k \triangleq \{x_k, x_{k-1}, \dots, x_o\} \tag{13.3.1}$$

and

$$J(X_k;k) \triangleq \min_{u_k,\dots,u_{N-1}} E_{|X_k} \exp\{\tfrac{1}{2}\sum_{i=k}^{N-1}(x_i'Q_ix_i+u_i'R_iu_i) + \tfrac{1}{2}x_N'Q_Nx_N\}, \tag{13.3.2}$$

where $E_{|X_k}$ denotes the expectation given X_k (i.e. the conditional expectation).

Using these definitions it is easy to deduce that

$$J(X_k;k) = \min_{u_k} [\exp\{\tfrac{1}{2}x_k'Q_kx_k + \tfrac{1}{2}u_k'R_ku_k\} \int_{-\infty}^{\infty} p(\omega_k;k)J(X_{k+1};k)d\omega_k], \tag{13.3.3}$$

where $p(\omega_k;k)$ is the probability density of ω_k. The boundary condition for (13.3.3) follows from (13.3.2) as

$$J(X_N;N) = \exp\{\tfrac{1}{2}x_N'Q_Nx_N\}. \tag{13.3.4}$$

Now, if $p(\omega_k;k)$ is a Gaussian probability density function, viz.

$$p(\omega_k;k) = \frac{1}{\sqrt{(2\pi)^q|\Lambda_k|}} \exp\{-\tfrac{1}{2}\omega_k'\Lambda_k^{-1}\omega_k\}, \tag{13.3.5}$$

it is easy to verify* by direct substitution that the solution of (13.3.3) is given by

$$J(X_k;k) = F_k \exp\{\tfrac{1}{2}x_k'P_k x_k\}, \quad k=0,\dots,N, \tag{13.3.6}$$

where, assuming that the necessary inverses exist and are positive definite,

$$P_k = Q_k + A_k'[\tilde{P}_{k+1} - \tilde{P}_{k+1}B_k(R_k + B_k'\tilde{P}_{k+1}B_k)^{-1}B_k'\tilde{P}_{k+1}]A_k \tag{13.3.7}$$

and

$$\tilde{P}_{k+1} = P_{k+1} + P_{k+1}\Gamma_k(\Lambda_k^{-1} - \Gamma_k'P_{k+1}\Gamma_k)^{-1}\Gamma_k'P_{k+1}, \tag{13.3.8}$$

and where

$$P_N = Q_N. \tag{13.3.9}$$

Furthermore,

$$F_k = F_{k+1} \frac{\sqrt{|(\Lambda_k^{-1} - \Gamma_k'P_{k+1}\Gamma_k)^{-1}|}}{|\Lambda_k|}, \quad F_N = 1 \tag{13.3.10}$$

and the optimal control policy is

$$u_k^* = -C_k x_k, \tag{13.3.11}$$

where

---

* The exponential nature of $p(\omega_k;k)$ allows the integration in (13.3.3) in closed form.

$$C_k = (R_k + B_k'\tilde{P}_{k+1}B_k)^{-1} B_k'\tilde{P}_{k+1}A_k, \quad k=0,\ldots,N-1. \qquad (13.3.12)$$

The above solution to the linear-exponential-Gaussian control problem has the following interesting properties.

(i) The optimal controller is a linear function of the system state.

(ii) The optimal controller depends upon the covariance of the system noise Λ_k, through (13.3.8).

(iii) The solution to the deterministic case may be obtained by letting $\Lambda_k \to 0$ in (13.3.8), which yields $\tilde{P}_k = P_k$, resulting in the same solution as given in (13.2.4) - (13.2.6).

(iv) If the covariance of the noise Λ_k is large enough, $(\Lambda_k^{-1} - \Gamma_k'P_{k+1}\Gamma_k)$ can become indefinite and (see [2]) this means that the optimal value of $E[\exp(\psi)]$ ceases to exist. Thus, noise that is 'too wild' causes the value of the exponential performance criterion to become unbounded.

Properties (ii) and (iv) distinguish the linear-exponential-Gaussian problem from the linear-quadratic-Gaussian case.

A similar formulation and solution are possible in continuous time, and the reader is referred to [2] for further details.

When the system state x_k cannot be measured directly and only a noisy observation (13.2.9) is available, the situation is much more compli= cated than in the linear-quadratic-Gaussian case. Here it turns out that the optimal controller is a linear function of $\hat{X}_k$ where

$$\hat{X}_k = E[X_k | y_k, y_{k-1}, \ldots, y_0]. \qquad (13.3.13)$$

In words, the optimal control at time k is a linear function of the mean values of all the vectors x_k, $x_{k-1},\dots,x_0$, conditional on all the observations $y_k, y_{k-1},\dots,y_0$. The number of components in the vector X_k increases by n at each time step and so the dimensions of the optimal gain matrices which determine the optimal control at each time step also increase .

When there is no process noise (i.e. $\Gamma_k\omega_k$ is not present) the situation is simpler, and the optimal control at each time step is a linear function of a 2n-vector. Another interesting special case occurs when $Q_k = 0$, $k=0,\dots,N-1$, when the optimal control is simply a linear function of the conditional mean $\hat{x}_k$, as in the linear-quadratic case. However, in this case the gain matrices still depend upon the covariances of the noises, in contrast to the linear-quadratic case.

13.4 NONLINEAR STOCHASTIC SYSTEMS

We now consider the case of minimizing $E[\psi]$, viz. the expected value of a quadratic performance criterion, subject to the nonlinear stochastic discrete-time dynamic system

$$x_{k+1} = A_k x_k + B_k u_k + f_k(x_k, u_k, \omega_k) \;, \quad x_o \text{ given} \,, \qquad (13.4.1)$$

where $\{\omega_k\}$ is a not necessarily Gaussian, independently distributed, zero mean random sequence. We assume that the vector x_k can be measured exactly.

Clearly (13.2.2) is a special case of (13.4.1) if f_k satisfies the following assumptions.

ASSUMPTION 13.4.1 $\bar{f}_k(x_k,u_k) \triangleq E[f_k(x_k,u_k,\omega_k)] = 0$ for all $x_k \in R^n$, $u_k \in R^r$, $k=0,\dots,N-1$.

The case of $\bar{f}_k(x_k,u_k)$ being a linear function of x_k and u_k presents no conceptual problems and is easily handled; only the resulting equations are more cumbersome [2].

ASSUMPTION 13.4.2 $F_k(x_k,u_k) \triangleq E[f_k(x_k,u_k,\omega_k)f_k'(x_k,u_k,\omega_k)]$ exists and is a homogeneous quadratic function of x_k,u_k, $k=0,\ldots,N-1$. That is, F_k has the representation

$$F_k(x_k,u_k) = T_k^0 + \sum_{i=1}^{n'} T_k^i(\tfrac{1}{2}x_k'W_k^i x_k + \tfrac{1}{2}u_k'M_k^i u_k) \qquad (13.4.2)$$

where $n' = n(n+1)/2$.

Naturally, because F_k is a covariance matrix, the parameters on the right-hand side of (13.4.2) must be such that $F_k(x_k,u_k)$ is positive semidefinite for all x_k,u_k. The case when F_k is a general quadratic form in x_k,u_k is treated in [2].

Under the above assumptions it turns out, using straightforward Dynamic Programming, that the optimal controller is given by

$$u_k^* = -\tilde{R}_k^{-1}\tilde{A}_k x_k \qquad (13.4.3)$$

where

$$\tilde{R}_k \triangleq R_k + B_k'P_{k+1}B_k + \bar{M}_k(P_{k+1}) \qquad (13.4.4)$$

$$\tilde{A}_k \triangleq B_k'P_{k+1}A_k \qquad (13.4.5)$$

$$\bar{M}_k(P_{k+1}) \triangleq \tfrac{1}{2}\sum_{i=1}^{n'} \mathrm{tr}(P_{k+1}T_k^i)M_k^i \qquad (13.4.6)$$

$$\bar{W}_k(P_{k+1}) \triangleq \tfrac{1}{2}\sum_{i=1}^{n'} \mathrm{tr}(P_{k+1}T_k^i)W_k^i \qquad (13.4.7)$$

$$P_k = Q_k + A_k'P_{k+1}A_k + \bar{W}_k(P_{k+1}) - \tilde{A}_k\tilde{R}_k^{-1}\tilde{A}_k, \quad P_N = Q_N \qquad (13.4.8)$$

$$e_k = \tfrac{1}{2}\,\mathrm{tr}(P_{k+1}\;T_k^o) + e_{k+1} \quad , \quad e_N = 0, \tag{13.4.9}$$

and the minimum value of $E[\psi]$ is given by

$$\tfrac{1}{2}\,x_o'\;P_o x_o + e_o. \tag{13.4.10}$$

The above result is interesting because of the linearity of the opti= mal feedback controller in the face of a *nonlinear* dynamic system. The feature not present in the linear-quadratic-stochastic case is the dependence of this linear controller on the covariance of the noise via (13.4.4), (13.4.6), (13.4.7) and (13.4.8).

The following special cases satisfy Assumptions 13.4.1-2, and accord= ingly give rise to the linear optimal controller (13.4.3) when $E[\psi]$ is minimized.

(i) $f_k(x_k,u_k,\omega_k) = \Gamma_k\omega_k$,
with

$$E[\omega_k] = 0, \quad E[\omega_k\omega_k'] = \Lambda_k.$$

This is the classical linear-quadratic-stochastic problem re= viewed in Section 13.2. Here,

$$F_k(x_k,u_k) = \Gamma_k\Lambda_k\Gamma_k'.$$

(ii) $f_k(x_k,u_k) = \|x_k\|\Gamma_k\omega_k$. This is the case of state-norm dependent noise, where

$$F_k(x_k,u_k) = x_k'x_k\Gamma_k\Lambda_k\Gamma_k'.$$

(iii) $f_k(x_k,u_k) = x_k\Gamma_k\omega_k$,

where $\Gamma_k' \in R^q$, $\omega_k \in R^q$.

This is the case of state multiplicative noise where

$$F_k(x_k,u_k) = x_k\Gamma_k\Lambda_k\Gamma_k'x_k'$$

$$= \Gamma_k\Lambda_k\Gamma_k'x_kx_k',$$

which can clearly be put in the form (13.4.2).

The extension to continuous time of the above general result has not appeared in the literature, although discussions of special cases such as state multiplicative noise are available.

The case of noisy measurements of x_k has not been properly explored, but it seems that a linear solution to the general problem does not exist.

13.5 REFERENCES

[1] J.S. Meditch, *Stochastic Optimal Linear Estimation and Control*. McGraw Hill, New York, 1969.

[2] D.H. Jacobson, *Extensions of Linear-Quadratic Control, Optimization and Matrix Theory*. Academic Press, New York, 1977.

14. DIFFERENTIAL GAMES

14.1 INTRODUCTION

Differential games can be viewed as obvious extensions of control pro= blems, namely as two-sided control problems. Thus, consider the system

$$\dot{x} = A(t)x + B_1(t)u + B_2(t)v, \quad 0 \leqslant t \leqslant T, \quad x(0)=x_0 \tag{14.1.1}$$

and the cost functionals

$$J_i(u(\cdot), v(\cdot)) = \int_0^T [u'R_i(t)u + x'Q_i(t)x]dt + x'(T)Q_{if}x(T), i=1,2. \tag{14.1.2}$$

Here the vectors $x, x_0 \in \mathbb{R}^n$, $u,v \in \mathbb{R}^m$ and A, $R_i, Q_i, Q_{if}, B_i, i=1,2$ are real piecewise continuous matrices with the proper dimensions; in addi= tion the matrices R_i, Q_i, Q_{if} $i=1,2$ are symmetric, and R_i, $i=1,2$, are positive definite. J_1 is the u-player's cost functional and J_2 is the v-player's cost functional. Both players, u and v, exercise control over the system trajectory $x(t)$, and their decisions (controls) jointly affect the cost functionals J_1 and J_2. It is the purpose of u and v to mini= mize their respective cost functionals J_1 and J_2.

However, there is more to differential games than specifying the dynamical (control) system (14.1.1) and the cost functionals (14.1.2). Specifical= ly, in Section 14.2 we shall dwell on the additional information required to give a full description of the differential game, and we shall give a brief survey of the problems which are germane to the study of differen= tial games. In Section 14.3 we present the solution for the Linear-Quad= ratic Differential Game (LQDG) and we also consider the open-loop solu= tions. In Section 14.4 we then present an example of an LQDG and in Section 14.5 we discuss the infinite-time LQDG.

14.2 THE DIFFERENTIAL GAME FORMULATION

14.2.1 Strategies

In (deterministic) optimal control theory and in the calculus of varia= tions the problem is posed of finding an element in a function space (a control function) $u(t), 0 \leqslant t \leqslant T$ which renders the cost functional optimal. In a differential game, a knowledge of the optimal controls $u(t)$, $v(t)$, $0 \leqslant t \leqslant T$, is sometimes (see last paragraph of Section 14.2.2) useful insofar as we are interested in the value(s) of the cost functional(s). (One could then decide whether the game is worth playing at all.)

If, however, the party is to take place, then strategies, and not optimal controls, are required. Indeed, since the system response is determined by the joint action of the two players, a strategy affords the players the opportunity to take instantaneous advantage of the adversary's errors.

In our LQDG setting it suffices to confine ourselves to feedback strate= gies, i.e. $u = F_1(t,x(t))$, $v = F_2(t,x(t))$ (although in general feedback controls are not the exclusive candidates for strategies in differential games). Thus, as far as the information pattern is concerned, we spe= cify that either player has access to the system state (more precisely, the system phase $(t,x(t))$).

14.2.2 The solution concept

DEFINITION *The pair of strategies* (u_N^*, v_N^*) *constitutes an (optimal) Nash equilibrium pair whenever*

$$J_1(u,v_N^*) \geqslant J_1(u_N^*,v_N^*)$$

and

$$J_2(u_N^*,v) \geqslant J_2(u_N^*,v_N^*)$$

for all strategies u,v *and subject to the differential constraint* (14.1.1).

DEFINITION *The pair of strategies* (u_p^*,v_p^*) *constitutes an (optimal) Pareto pair whenever for some* u,v *and subject to the differential con= straint* (14.1.1),

$$J_i(u,v) \leqslant J_i(u_p^*,v_p^*) \text{ for } i=1,2$$

implies that

$$J_i(u,v) = J_i(u_p^*,v_p^*) \text{ for } i=1,2.$$

Note that both players could do better if they would adhere to the Pareto optimum; however, then there is no guarantee against cheating by the adversary, as is the case with the Nash optimal point, where it does not pay either player to deviate unilaterally from the equili= brium point u_N,v_N.[1)] In fact, the Nash and Pareto concepts of optimality are not the only ones, and the analyst has to choose the concept best suited to the problem under consideration.

If $J \triangleq J_1 = -J_2$, we have the important special case of a zero-sum LQDG. Thus we have the following definition.

DEFINITION *The pair of strategies* u*, v* *is a saddle point if for all strategies* u *and* v,

---

1) This is the 'prisoner's' (or 'arms' limitations) dilemma.

$$\max_v \min_u J(u,v) = \min_u \max_v J(u,v) = J(u^*,v^*). \qquad (14.2.1)$$

Obviously, in the zero-sum case a saddle point is equivalent to a Nash equilibrium point. Specifically, in order to establish the existence of a saddle point one must show that

$$\max_v \min_u J(u,v) \geqslant \min_u \max_v J(u,v);$$

the converse inequality holds under relatively mild assumptions. Assume now that the max-min is attained at $\bar{u}$, $\bar{v}$, so that evidently

$$\max_v \min_u J(u,v) \leqslant J(u,\bar{v}) \text{ for all u strategies.} \qquad (14.2.2)$$

Similarly, if the min-max is attained at $\bar{\bar{u}}$, $\bar{\bar{v}}$, we have that

$$\min_u \max_v J(u,v) \geqslant J(\bar{\bar{u}},v) \text{ for all v strategies.} \qquad (14.2.3)$$

We therefore realize that establishing the existence of a saddle point provides us with a 'minimum performance' guarantee, viz. the relations (14.2.1) - (14.2.3) amount to:

$$J(u^*,v) \leqslant J(u^*,v^*) \leqslant J(u,v^*) \qquad (14.2.4)$$

for all strategies u and v.

Furthermore, in view of (14.2.4) it is readily verifiable that the saddle point value $J(u^*,v^*)$ is unique. Note however that (14.2.4) does not imply that lower (higher) values than $J(u^*,v^*)$ for the cost functional cannot be attained; in fact if the minimizer (maximizer) has a priori knowledge of the v (u)strategy, then lower (higher) values than $J(u^*,v^*)$ for the cost functional could be realized.

Finally, if we have an ('open-loop') saddle point in control functions u(t),v(t), $0 \leq t \leq T$ (we then refer to 'a pure strategy'), we can readily compute the value of the saddle point. However, this is somewhat exceptional, and as a rule

$$\max_{\substack{v(t)u(t)\\ 0 \leq t \leq T}} \min J(u(\cdot),v(\cdot)) < \min_{u(t)v(t)} \max J(u(\cdot),v(\cdot)).$$

Bridging this gap is the subject of the next subsection.

14.2.3 The game-theoretic aspect

It has been shown by Von Neumann and Morgenstern [1] that in (discrete) matrix games, in order to bridge the maximin-minimax gap, namely to achieve a saddle point, and therefore have the guarantee given by (14.2.4), one must have recourse to 'mixed strategies', which amounts to specifying probability distributions for each player, according to which either player is then going to choose his row or column entry.

It has been shown by Elliot, Kalton and Markus [2] that in the specific differential game setting, for systems and payoff functionals which are linear in the state variables, one can achieve a saddle point using so-called 'relaxed' controls. Moreover, if in addition the Hamiltonian of the problem does not have cross coupling terms in u and v, one can prove the existence of saddle points in chattering controls.

A different approach is taken by Fleming [3], Varayia-Lin [4] and Friedman [5]. They partition the time interval [0,T] into a set of n=1,2,... subintervals and consider two sets of optimal control problems:

$$V_n \triangleq \max_{\tilde{v}_1} \min_{\tilde{u}_1} \max_{\tilde{v}_2} \min_{\tilde{u}_2} \ldots\ldots \max_{\tilde{v}_n} \min_{\tilde{u}_n} J(u,v)$$

and

$$V^n \triangleq \min_{\tilde{u}_1} \max_{\tilde{v}_1} \min_{\tilde{u}_2} \max_{\tilde{v}_2} \dots \min_{\tilde{u}_n} \max_{\tilde{v}_n} J(u,v);$$

here $\tilde{u}_i$, $\tilde{v}_i$ are control functions defined on the i-th subinterval (of length T/n), i=1,2,...,n and $u = \tilde{u}_1 . \tilde{u}_2 \dots . \tilde{u}_n$, $v = \tilde{v}_1 . \tilde{v}_2 . \dots . \tilde{v}_n$ (the symbol • stands for concatenation); V^n and V_n are referred to as upper and lower values respectively.

It is shown that $V^n \geqslant V_n \; \forall \; n = 1,2,\dots$ and conditions are given for

$$\lim_{n \to \infty} V^n = \lim_{n \to \infty} V_n = V.$$

Indeed, this is the case in the zero-sum LQDG.

14.2.4 Specific aspects of solving differential games

As we have emphasized in Section 14.2.2, in DGs we seek feedback (closed-loop) control laws $u(t,x(t))$, $v(t,x(t))$ rather than (open-loop) optimal controls $u(t)$, $v(t)$, $0 \leqslant t \leqslant T$. Therefore the DG optimization problem does not fit well into the calculus of variations setting or, for that matter, into the functional analytic framework, where one seeks an element $u^*(t)$, (or $v^*(t)$) $0 \leqslant t \leqslant T$, in function space. In addition, even if we should be interested in the optimal controls (see the last paragraph of Section 14.2.2), we realize that in a DG setting the calculus of variations formalism is quite weak(insofar as characterizing the optimal controls is concerned).

Indeed, the first-order necessary conditions do not distinguish between the two-sided optimal control problems $\max_{v(t)} \min_{u(t)} J(u,v)$ and $\min_{u(t)} \max_{v(t)} J(u,v)$ and certainly, do not furnish the saddle point equality (14.2.1).

Furthermore, numerical methods which are suggested by an approach of linearization (e.g. gradient-type methods) are in general not applic= able in a DG setting. Indeed, it may be unprofitable for the adversary to adhere to a small perturbation of his control and therefore the deviation from the nominal system trajectory could be large, which in turn invalidates linearization [6].

Therefore, in order to derive a solution in DG one generally employs the global Hamilton-Jacobi-dynamic programming/regular synthesis ap= proach [7] which provides sufficient conditions for the construction of a solution. Loosely speaking, this amounts to solving a nonlinear Hamilton-Jacobi partial differential equation and verifying that the (optimal) flow field synthesized in this way makes sense. The trouble though is that it is necessary to allow for singularities of the solu= tion (discontinuities) in the domain of definition of the partial dif= ferential equation, and it is therefore necessary in fact to solve a set of Hamiltonian-Jacobi partial differential equations coupled along the surfaces of discontinuity, and patch the solutions together.

Fortunately, in LQDG this only amounts to solving a matrix Riccati differential equation.

14.3 THE FINITE-TIME ZERO-SUM LINEAR-QUADRATIC DIFFERENTIAL GAME

The cost functional is now given by

$$J(u(\cdot),v(\cdot)) = \int_0^T [u'R_1(t)u-v'R_2(t)v + x'Q(t)x]\,dt + x'(T)Q_f x(T); \tag{14.3.1}$$

where Q and Q_f as real symmetric $n\times n$ matrices, and in addition $Q(\cdot)$ is piecewise continuous.

The optimal feedback strategies are given by

$$u^*(t,x(t)) = -R_1^{-1}(t)B_1'(t)P(t)x(t) \qquad (14.3.2)$$

$$v^*(t,x(t)) = R_2^{-1}(t)B_2'(t)P(t)x(t) \qquad (14.3.3)$$

where $P(t)$, $0 \leqslant t \leqslant T$, is the solution to the matrix Riccati differen= tial equation

$$-\dot{P} = A'(t)P + PA(t) - P(B_1(t)R_1^{-1}(t)B_1'(t)-B_2(t)R_2^{-1}(t)B_2'(t))P + Q(t), \qquad (14.3.4)$$

with the end-point condition

$$P(T) = Q_f. \qquad (14.3.5)$$

In addition

$$J(u^*(\cdot),v^*(\cdot)) = x_0'P(0)x_0. \qquad (14.3.6)$$

We have tacitly assumed herein that there exists a solution on $[0,T]$ to the differential equations (14.3.4), (14.3.5) (i.e. T is less than the finite escape time of (14.3.4), (14.3.5)).

It is also of interest to consider the following two optimal control problems:

P1 : $\max\limits_{v(\cdot)} J(u(\cdot),v(\cdot))$,

where J is given by (14.3.1), subject to (14.1.1) and (14.3.2) in which $P(t)$ is given by (14.3.4),(14.3.5);

and

P2 : $\min\limits_{u(\cdot)} J(u(\cdot),v(\cdot))$,

where J is given by (14.3.1), subject to (14.1.1) and (14.3.3), in which $P(t)$ is given by (14.3.4), (14.3.5).

Both P1 and P2 are standard (time-varying) LQ optimal control problems, so that for problem P1 we find that

$$v = + R_2^{-1}(t)B_2'(t)P^{(1)}(t)x(t), \qquad (14.3.7)$$

where

$$-\dot{P}^{(1)} = (A(t)-B_1(t)R_1^{-1}(t)B_1'(t)P(t))'P^{(1)} + P^{(1)}(A(t)-B_1(t)R_1^{-1}(t)B_1'(t)P(t)) + P^{(1)}B_2(t)R_2^{-1}(t)B_2'(t)P^{(1)} + Q(t) + P(t)B_1(t)R_1^{-1}(t)B_1'(t)P(t), \qquad (14.3.8)$$

$$P^{(1)}(T) = Q_f, \qquad (14.3.9)$$

and for P2 we have that

$$u = -R_1^{-1}(t)B_1'(t)P^{(2)}(t)x(t), \qquad (14.3.10)$$

where

$$-\dot{P}^{(2)} = (A(t)+B_2(t)R_2^{-1}(t)B_2'(t)P(t))'P^{(2)} + P^{(2)}(A(t)+B_2(t)R_2^{-1}(t)B_2'(t)P(t)) - P^{(2)}B_1(t)R_1^{-1}(t)B_1'(t)P^{(2)} + Q(t) - P(t)B_2(t)R_2^{-1}(t)B_2(t)P(t), \qquad (14.3.11)$$

$$P^{(2)}(T) = Q_f. \qquad (14.3.12)$$

In view of (14.3.4), (14.3.5), (14.3.8), (14.3.9), (14.3.11) and (14.3.12) we conclude that $P^{(1)}(t) = P^{(2)}(t) = P(t)$ on $[0,T]$ so that the feed=back controls given by (14.3.2) and (14.3.10) for the u player (and similarly (14.3.3) and (14.3.7) for the v player) are identical.

If however, we try to solve the two-sided optimal control problem (14.1.1) and (14.3.1) for the optimal controls $u^0(t)$, $v^0(t) \in L^2[0,T]$ we find that

$$\max_{v(\cdot)} \min_{u(\cdot)} J(y(\cdot),v(\cdot)) < \min_{u(\cdot)} \max_{v(\cdot)} J(u(\cdot),v(\cdot)).$$

Specifically consider the optimal control problem

$$P3 \quad : \qquad \max_{v(\cdot)} J(u(\cdot),v(\cdot)),$$

where J is given by

$$J(u(\cdot),v(\cdot)) = \int_0^T [u'R_1(t)u - v'R_2(t)v]\,dt + x'(T)Q_f x(T), \qquad (14.3.13)$$

(i.e. we assume for the sake of simplicity that $Q(t) = 0$)

subject to (14.1.1) and

$$u^0(t) = -R_1^{-1}(t)B_1'(t)\Phi'(0,t)P(0)x_0; \qquad (14.3.14)$$

here $\Phi(t,\tau)$ is the fundamental matrix for the homogeneous system

$$\dot{x} = A(t)x. \qquad (14.3.15)$$

In (14.3.14) $u^0(t)$ is thus the open-loop optimal control function which corresponds to the optimal feedback control (14.3.2) if we bear in mind that the adjoint variable $\lambda(t)$ satisfies

$$\lambda(t) = P(t)x(t),$$

$$u^0(t) = -R_1^{-1}B_1'(t)\lambda(t),$$

and (since $Q = 0$)

$$\lambda(t) = \Phi'(0,t)\,\lambda(0).$$

P3 is therefore an optimal control LQ problem with a given disturbance function $u^0(t)$ in (14.1.1) given by (14.3.14). The solution to the LQ optimal control problem P3 is given by

$$v(t) = R_2^{-1}(t)B_2'(t)\,[P^{(3)}(t)x(t)+a(t)]$$

where

$$-\dot{P}^{(3)} = A'(t)P^{(3)} + P^{(3)}A(t) + P^{(3)}B_2(t)R_2^{-1}(t)B_2'(t)P^{(3)}, \quad (14.3.16)$$

$$P^{(3)}(T) = Q_f, \quad (14.3.17)$$

$$\dot{a} = -A'(t)a - P^{(3)}B_2(t)R^{-1}_2(t)B_2'(t)a - P^{(3)}B_1(t)u^0(t), \quad (14.3.18)$$

$$a(T) = 0. \quad (14.3.19)$$

Similarly, we pose the LQ optimal control problem

$$P4 \quad : \quad \min_{u(\cdot)} J(u(\cdot),v(\cdot)),$$

where J is given by (14.3.13), subject to (14.1.1) and the disturbance function $v^0(t)$

$$v^0(t) = R_2^{-1}(t)B_2'(t)\Phi'(0,t)P(0)x_0. \quad (14.3.20)$$

The solution to the LQ optimal control problem P4 is given by

$$u(t) = -R^{-1}_1(t)B_1'(t)\,[P^{(4)}(t)\,x(t) + b(t)],$$

where

$$-\dot{P}^{(4)} = A'(t)P^{(4)} + P^{(4)}A(t) - P^{(4)}B_1(t)R_1^{-1}(t)B_1'(t)P^{(4)}, \quad (14.3.21)$$

$$P^{(4)}(T) = Q, \quad (14.3.22)$$

and

$$\dot{b} = A'(t)b + P^{(4)}B_1(t)R_1^{-1}(t)B_1'(t)b + P^{(4)}B_2(t)v^0(t), \quad (14.3.23)$$

$$b(T) = 0. \quad (14.3.24)$$

Note that $P^{(3)} \neq P^{(1)} = P$ and $P^{(4)} \neq P^{(2)} = P$!

14.4. AN EXAMPLE

The equations of motion for an interceptor (pursuer) p and an evader e [8] are

$$\dot{v}_p = g + a_p \quad , \quad \dot{v}_e = g + a_e,$$

$$\dot{r}_p = v_p \quad , \quad \dot{r}_e = v_e \quad ,$$

where

v is the velocity vector of a body in three dimensions;

r is the position vector of a body in three dimensions;

g is the gravitational force vector per unit mass exerted on the body;

a is the control acceleration vector of a body.

The cost functional is

$$J = \alpha \, \| r_p(T) - r_e(T) \|_2^2 + \int_0^T [c_p^{-1} \| a_p \|_2^2 - c_e^{-1} \| a_e \|_2^2] \, dt,$$

where the constants α, c_p, c_e are positive. This means that the pursuer would like to decrease the miss distance $\| r_p(T) - r_e(T) \|_2$ without spending too much control energy $\int_0^T \| a_p \|_2^2 dt$; the evader would like to see to it that the miss distance and the pursuer control effort are large, whereas his own control effort is not too large. Applying to this pursuit-evasion model the results of Section 14.3, viz. equations (14.3.2) and (14.3.3), we obtain the following optimal feedback strategies:

$$a_p = \frac{-c_p(T-t)\,[r_p-r_e + (v_p-v_e)(T-t)]}{(1/\alpha)+(c_p-c_e)[(T-t)^3/3]} \qquad (14.4.1)$$

and

$$a_e = \frac{c_e}{c_p}\, a_p. \qquad (14.4.2)$$

Assuming that $c_p > c_e$ and letting $\alpha \to \infty$ (i.e. insisting on a small miss distance), the control strategy for the pursuer is then

$$a_p = -\frac{3}{(1-c_e/c_p)} \cdot \frac{r_p-r_e + (v_p-v_e)(T-t)}{(T-t)^2}. \qquad (14.4.3)$$

Furthermore, if the pursuer and the evader are on a nominal collision course and the nominal range is

$$R(t) = V(T-t),$$

where V is the nominal closing speed, and if x_p and x_e denote the lateral deviations (from the nominal collision course) of the pursuer and the evader respectively, then the geometry is as sketched below.

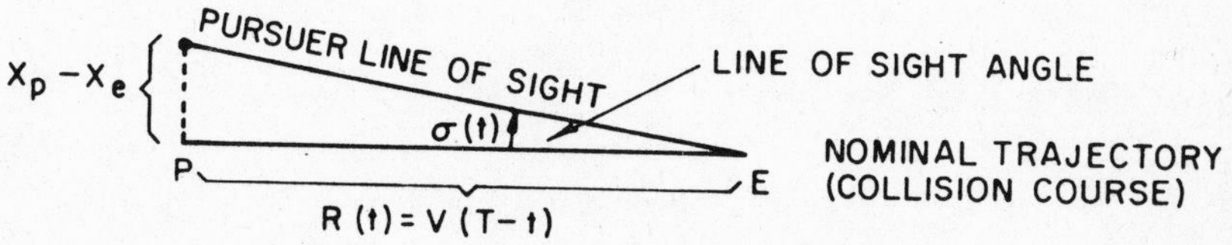

Basically, we assume here that all the maneuvering is done in a plane which is perpendicular to the nominal (initial) collision course.

Then since

$$x_p - x_e \approx R(t)\sigma = v(T-t)\sigma \qquad (14.4.4)$$

and

$$v_p = \dot{x}_p \ , \quad v_e = \dot{x}_e \ ,$$

$$v_p - v_e = \frac{d}{dt}(x_p - x_e) = V(T-t)\dot{\sigma} - V\sigma \ , \tag{14.4.5}$$

we have, upon substituting (14.4.4) and (14.4.5) in (14.4.2), that the pursuer feedback control (lateral acceleration) is

$$a_p = \frac{3}{(1-c_e/c_p)} \quad V\dot{\sigma}$$

which is the well known proportional navigation guidance law with the navigation constant $K = \frac{3}{1-c_e/c_p}$.

In practice $3 \leqslant K \leqslant 5$. Note that the value of K=3 corresponds to $c_e = 0$, i.e. an evader which is completely unmaneuverable.

14.5 INFINITE TIME LINEAR-QUADRATIC DIFFERENTIAL GAMES

We shall assume in this section that the matrices A, B_1, B_2, R_1, R_2 and Q are time-invariant, and in addition that the matrices $Q_f = 0$ and Q are nonnegative definite, viz. $Q = H'H$ for some m×n matrix H.

Let $T \to \infty$, where T is the duration of the game, and assume that the solution to the differential matrix Riccati equation (14.3.4), (14.3.5) P(0;T) converges, i.e.

$$\lim_{T \to \infty} P(0;T) = P. \tag{14.5.1}$$

Then, in line with the well-known results from optimal control theory for the linear-quadratic regulator, it is plausible to conjecture that the value of the game, J_∞, would then be

$$J_\infty(x_0) = x_0'Px_0$$

and (see (14.3.2) and (14.3.3)):

$$u^*(x(t)) = -R_1^{-1}B_1'Px(t), \tag{14.5.2}$$

$$v^*(x(t)) = R_2^{-1}B_2'Px(t). \tag{14.5.3}$$

It is surprising that this conjecture is not true [9].

EXAMPLE 14.5.1

Consider the scalar game with the following parameters

$$A = B_1 = B_2 = Q = R_1 = 1,\ Q_f = 0,\ R_2 = 2.$$

The Riccati equation (14.3.4), (14.3.5) is now

$$\dot{p} = \tfrac{1}{2}p^2 - 2p-1,$$

$$p(T) = 0,$$

which can be integrated to give

$$p(t;T) = \sqrt{6}\,\tanh[\sqrt{6}(T-t)+\tanh^{-1}(-2/\sqrt{6})]+2.$$

Hence,

$$\lim_{T\to\infty} p(0;T) = 2 + \sqrt{6}$$

so that the strategies in (14.5.2) and (14.5.3) are

$$u^*(x(t)) = -(2 + \sqrt{6})x(t) \tag{14.5.4}$$

$$v^*(x(t)) = \frac{2+\sqrt{6}}{2}\,x(t). \tag{14.5.5}$$

These strategies are not in equilibrium. Indeed, consider the problem P2 : the minimizer is then faced with the optimal control problem

$$\tilde{J}_\infty(x\cdot) \triangleq \min_u \int_0^\infty [u^2 - (4+2\sqrt{6})x^2]\,dt$$

subject to

$$\dot{x} = \frac{4+\sqrt{6}}{2}\, x + u \quad , \quad x(0) = x_0 .$$

Now the minimizer's optimal response is not given by (14.5.4).

In fact, by playing

$$u(x) = 0$$

the minimizer can obtain a value

$$\tilde{J}_\infty\,(x_0) = -\infty \text{ provided } x_0 \neq 0.$$

Indeed, the following can be shown to hold [9], [10]: if there exists a real symmetric nonnegative definite solution to the algebraic matrix Riccati equation

$$A'P + PA - P(B_1R_1^{-1}B_1' - B_2R_2^{-1}B_2')P + Q = 0 \qquad (14.5.6)$$

then there exists a symmetric nonnegative definite solution P of (14.5.6) such that the value of the (infinite time) game is $x_0'Px_0$; P is (in the sense of definite matrices) the smallest symmetric nonnegative definite solution of (14.5.6). Here we mean by 'value ' a number p is the *value* of the (infinite time) game if, for any $\varepsilon > 0$, there exist strategies u_ε, v_ε, such that

$$J(u_\varepsilon, v) \leqslant p + \varepsilon \text{ for all } v \in V$$

and

$$J(u, v_\varepsilon) \geqslant p - \varepsilon \text{ for all } u \in U,$$

where we denote by U and V the sets of permissible strategies for the minimizing and maximizing players respectively. The strategies u_ε and v_ε are called ε-optimal.

The strategies given by (14.3.2) and (14.3.3) do not in general form an equilibrium pair. Moreover, the null space of the matrix P is equal to the set of all unobservable states in the open-loop system (A,H). In addition, if an initial state x_0 is unobservable in the un= controlled system (A,H) then that state is also unobservable in the system

$$\dot{x} = [A-(B_1R_1^{-1}B_1'-B_2R_2^{-1}B_2')P]\,x$$

with output

$$y = Hx.$$

Furthermore, if the pair (A,H) is observable then we have for the 'closed-loop' system

$$Re(\lambda([A-(B_1R_1^{-1}B_1' - B_2R_2^{-1}B_2')P])) \leqslant 0.$$

Alternatively [11], if we assume that there exists a real symmetric solution $\bar{P}$ of equation (14.5.6) which is strictly feedback-stabilizing, i.e.

$$Re(\lambda([A-(B_1R_1^{-1}B_1'-B_2R_2^{-1}B_2')\bar{P}])) < 0,$$

then the strategies given by (14.5.2) and (14.5.3) form an equilibrium pair for the infinite-time problem in the class of admissible strategy pairs UV, where $UV \triangleq \{u(x,t),v(x,t)|$ the solution of $\dot{x} = Ax + B_1u(x,t) + B_2v(x,t)$ is defined for all initial conditions x_0, and is such that $x(t) \to 0$ as $t \to \infty\}$.

14.6 REFERENCES

[1] John von Neumann and Oskar Morgenstern, *Theory of Games and Economic Behaviour.* Princeton University Press, 1947.

[2] R.J.Elliot, N.J. Kalton, L. Markus, Saddle points for linear differential games. *SIAM J. Control,* Vol 11, No 1, 1973,pp 100-112.

[3] W.H. Fleming, The convergence problem for differential games. *J. Math. Analysis Appl.*, Vol 13, 1961, pp 102-116.

[4] P. Varaiya, J. Lin, Existence of saddle points in differential games. *SIAM J. Control,* Vol 7, 1969, pp 142-157.

[5] A. Friedman, *Differential Games.* Wiley-Interscience, 1971.

[6] G.L. Olsder, Comments on a numerical procedure for the solution of differential games. *IEEE Trans. on Aut.Contr.*, Vol AC-14, 1975, p 64.

[7] W.H. Fleming, R.W. Rishel, *Deterministic and Stochastic Optimal Control.* Springer-Verlag, New York, 1975.

[8] Y.C. Ho, A.E. Bryson, S. Baron, Differential games and optimal pursuit-evasion strategies. *IEEE Trans. on Aut.Contr.*, Vol AC-10, No 4, Oct 1965, pp 385-389.

[9] E.F. Mageirou, Values and strategies for infinite-time linear-quadratic games. *IEEE Trans. on Aut. Contr.*, Vol AC-21, 1976, pp 547-550.

[10] M.Pachter, Some properties of the value matrix in infinite-time linear-quadratic differential games. *IEEE Trans. on Aut. Contr.*, Vol AC-23, 1978, pp 746-748.

[11] D.H. Jacobson, On values and strategies for infinite-time linear quadratic games. *IEEE Trans. on Aut. Contr.*, Vol AC-22, 1977, pp 490-491.

Additional bibliography on controllability and feedback in linear differential games:

Heymann, M., Pachter, M., Stern, R.J., Max-min control problems: a system theoretic approach. *IEEE Trans. on Aut. Control*, Vol AC-21, 1976, pp 455-463.

Heymann, M., Pachter, M., Stern, R.J., Weak and strong max-min control=lability. *IEEE Trans. on Aut. Control*, Vol AC-21, No 4, Aug 1976, pp 612-613.

Heymann, M., Pachter, M., Stern, R.J., On linear games with subspace target. *Funkcialaj Ekvacioj*, Vol 20, No 1, April 1977, pp 71-76.

Heymann, M., Pachter, M., On the max-min pursuit-game. *J. of Math. Analysis and Applic.*, Vol 70, No 2, Aug 1978, pp 430-444.

To the best of our knowledge, the following is a complete list of monographs on differential games.

Isaacs, R., *Differential Games*. Wiley, New York, 1965.

Blaquierè, A., Gérard, F; Leitmann, G., *Quantitative and Qualitative Games*. Academic Press, 1969.

Hājek, O., *Pursuit Games*. Academic Press, 1975.

Krasovskii, N., Pontryagin, L.S., *Differential Games*. (1977)(in russian).

Kuhn, H.W. and Szegö, G., eds., *Differential Games and Related Topics*. Elsevier, 1971.

Danskin, J.M., *Value in Differential Games*. Unpublished manuscript.

Proceedings of the conference on *The Theory and Application of Differen=tial Games*. 27 Aug-6 Sept. 1974, Cont. Theory Center, University of Warwick, 1974.

Roxin, E.O., Pan Tai Liu, R.L. Steinberg, eds. *Differential games and control theory*, Proceedings of the Kingston Conferences held at the University of Rhode Island (Vols 1,2,3), published by Marcel Dekker, 1974, 1976, 1978.

Olsder, G.M. (ed.), *Differential Games and Applications*. Proceedings of a Workshop, Enschede, 1977, Springer, 1977.

15. OPTIMAL CONTROL OF PARTIAL DIFFERENTIAL EQUATIONS

15.1 AN EXAMPLE OF BOUNDARY CONTROL

The study of different sorts of physical phenomena leads to various dif= ferent types of partial differential equations, and their mathematical treatment is accordingly specialized. Here we shall restrict ourselves to the examination of a specific control problem where the system is governed by the heat equation, and try to illustrate certain aspects of the theory which are not immediate extensions of the control theory of linear ordinary differential equations.

Let Ω be a bounded domain in $\mathbb{R}^n$ with sufficiently smooth boundary Γ. Let $T > 0$ be fixed and let $Q = \Omega \times (0,T)$, $\Sigma = \Gamma \times (0,T)$. The controls v are assumed to belong to U_{ad}, the set of admissible controls, U_{ad} being a closed, convex subset of $L^2(\Sigma)$. For each $v \in U_{ad}$, the state $y(\cdot,\cdot;v)$ is determined by the mixed initial-boundary-value problem

$$\frac{\partial}{\partial t} y(x,t;v) - \Delta y(x,t;v) = 0, \quad (x,t) \in Q,$$

$$\frac{\partial y}{\partial \nu}(x,t;v) = v(x,t), \; (x,t) \in \Sigma, \qquad (15.1.1)$$

$$y(x,0;v) = \overset{0}{y}(x), \quad x \in \Omega,$$

where $\Delta = \sum_{i=1}^{n} \frac{\partial^2}{\partial x_i^2}$ is the Laplacian, $\frac{\partial}{\partial \nu}$ denotes differentiation in the direction of the unit normal vector at a boundary point of Ω, pointing to the exterior of Ω, and $\overset{0}{y}$ is a given function in $L^2(\Omega)$.

We shall consider the optimization problem

(P) $$\inf_{v \in U_{ad}} J(v),$$

where

$$J(v) = \int_0^T \int_\Omega (y(x,t;v) - y_d(x,t))^2 dxdt + \int_0^T \int_\Gamma v^2(x,t) d\Gamma dt, \tag{15.1.2}$$

with a given function $y_d \in L^2(Q)$ ($d\Gamma$ denotes the element of area on Γ).

In Section 15.2 we shall introduce manageable notation, and define what is meant by a solution of (15.1.1) for $v \in L^2(\Sigma)$. In Section 15.3 we shall derive a characterization of the optimal control u by introducing the adjoint state. In Section 15.4 we shall outline the derivation of the feedback form of the solution in the unconstrained case ($U_{ad} = L^2(\Sigma)$), and in Section 15.5 we shall cite references concerning the semigroup-theoretic approach and controllability.

15.2 THE GENERALIZED SOLUTIONS OF THE EQUATION

Our treatment will follow Lions [6].

For $t \in [0,T]$ and $v \in L^2(\Sigma)$, define $v(t)(\cdot)$ by

$$v(t)(x) = v(x,t) \text{ for } x \in \Gamma.$$

Let $y_v(t)(\cdot)$ be defined by

$$y_v(t)(x) = y(x,t;v) \quad , \; t \in [0,T], \; x \in \Omega.$$

Thus, $y_v: [0,T] \to L^2(\Omega)$, $v: [0,T] \to L^2(\Gamma)$, and similarly, $\frac{\partial}{\partial\nu} y_v : [0,T] \to L^2(\Gamma)$ are Hilbert space-valued functions. Equation (15.1.1) can be written as

$$\begin{aligned} \frac{d}{dt} y_v(t) - \Delta y_v(t) &= 0, \quad t \in (0,T), \\ \frac{\partial}{\partial\nu} y_v(t) &= v(t), \quad t \in (0,T), \\ y_v(0) &= \overset{0}{y} . \end{aligned} \tag{15.2.1}$$

Let $(y,z)_\Omega$ denote the inner product in $L^2(\Omega)$:

$$(y,z)_\Omega = \int_\Omega y(x)z(x)dx \qquad (\|y\|_\Omega = \sqrt{(y,y)_\Omega}).$$

Similarly,

$$(v,w)_\Gamma = \int_\Gamma v(x)w(x)d\Gamma$$

denotes the inner product in $L^2(\Gamma)$;

$$(y,z)_Q = \int_0^T (y(t),z(t))_\Omega dt$$

denotes the inner product in $L^2(Q)$ $(=L^2(0,T;L^2(\Omega)))$;

and

$$(v,w)_\Sigma = \int_0^T (v(t),w(t))_\Gamma dt$$

denotes the inner product in $L^2(\Sigma)(=L^2(0,T;L^2(\Gamma)))$.

With this notation we can write

$$J(v) = \|y_v - y_d\|_Q^2 + \|v\|_\Sigma^2. \tag{15.2.2}$$

If y_v is a classical solution of (15.2.1), then for each sufficiently smooth ψ defined in Ω,

$$\frac{d}{dt}(y_v(t),\psi)_\Omega - (\Delta y_v(t),\psi)_\Omega = 0,$$

and applying Green's formula,

$$\frac{d}{dt}(y_v(t),\psi)_\Omega - (v(t),\gamma_0\psi)_\Gamma + a(y_v(t),\psi) = 0, \tag{15.2.3}$$

where $\gamma_0\psi$ is the restriction of ψ to Γ, and

$$a(y,\psi) = \int_\Omega \sum_{i=1}^n \frac{\partial y}{\partial x_i}\frac{\partial \psi}{\partial x_i}dx.$$

This gives us the clue as to how a generalized solution for (15.2.1) may be defined.

Let us introduce the Sobolev space

$$H^1(\Omega) = \{f \in L^2(\Omega) : \frac{\partial f}{\partial x_i} \in L^2(\Omega), \quad i=1,2,\ldots,n\}.$$

For any $f \in H^1(\Omega)$, the trace $\gamma_0 f$ on the boundary Γ makes sense and ex= tends the restriction mapping $f \longmapsto f|_\Gamma$ for functions smooth on $\bar{\Omega}$. The generalized solution y_v of (15.2.1) will be a function $y_v \in L^2(0,T;H^1(\Omega))$ such that (15.2.3) is satisfied for all $\psi \in H^1(\Omega)$, a.e. on $(0,T)$, and $y_v(0) = \overset{o}{y}$. Here $\frac{d}{dt}(y_v(t),\psi)_\Omega$ is in the sense of distributions on $(0,T)$. It is known [6] that such a generalized solution exists, that it is unique, and is continuous on $[0,T]$ as an $L^2(\Omega)$-valued function. The initial value $\overset{o}{y}$ is attained at $t=0$ in the sense that

$$\lim_{t \downarrow 0} \|y_v(t) - \overset{o}{y}\|_\Omega = 0.$$

15.3 CONDITIONS FOR OPTIMALITY

THEOREM 15.3.1 *The problem*

$$\text{(P)} \qquad \inf_{v \in U_{ad}} J(v)$$

has the unique solution $u \in U_{ad}$ *characterized by*

$$(\gamma_0 p_u + u, v-u)_\Sigma \geq 0, \; v \in U_{ad}, \tag{15.3.1}$$

where

$$\begin{aligned} &\frac{d}{dt} y_u - \Delta y_u = 0 \text{ in } Q, \\ &\frac{\partial}{\partial \nu} y_u = u \text{ on } \Sigma, \\ &y_u(0) = \overset{o}{y}, \end{aligned} \tag{15.3.2}$$

in the sense of generalized solutions described in Section 15.2, *and*

$$-\frac{d}{dt}p_u - \Delta p_u = y_u - y_d \text{ in } Q,$$

$$\frac{\partial p_u}{\partial \nu} = 0 \text{ on } \Sigma, \qquad (15.3.3)$$

$$p_u(T) = 0,$$

in the sense that $p_u \in L^2(0,T; H^1(\Omega))$, $p_u(T) = 0$, *and*

$$-\frac{d}{dt}(p_u(t),\psi)_\Omega + a(p_u(t),\psi) = (y_u(t)-y_d(t),\psi)_\Omega \qquad (15.3.4)$$

for each $\psi \in H^1(\Omega)$ *and a.e. on* $(0,T)$.

PROOF It is observed that $L : L^2(\Sigma) \to L^2(Q)$ defined by $Lv = y_v - y_0$ (y_0 being the state corresponding to the control which is identically zero) is bounded and linear, so that

$$J(v) = \|Lv - (y_d-y_0)\|_Q^2 + \|v\|_\Sigma^2$$

is a continuous, coercive ($J(v) \geqslant \|v\|_\Sigma^2$) quadratic functional on $L^2(\Sigma)$. Thus there exists a unique solution $u \in U_{ad}$ of (P), and u is characterized by

$$\langle J'(u), v-u\rangle \geqslant 0, \qquad v \in U_{ad},$$

where $J' : L^2(\Sigma) \to L^2(\Sigma)'$ is the Gateaux derivative of J [6]. It is easily computed that

$$\langle J'(u),v\rangle = 2(y_u-y_d,Lv)_Q + 2(u,v)_\Sigma,$$

so that $u \in U_{ad}$ is optimal iff

$$(y_u-y_d, L(v-u))_Q + (u,v-u)_\Sigma \geqslant 0 \qquad (15.3.5)$$

for all $v \in U_{ad}$.

The optimality condition (15.3.5) is in terms of Lv, Lu, and we should like to express the condition directly in terms of v,u. For this purpose we introduce $L^*: L^2(Q) \to L^2(\Sigma)$, the adjoint of L, and write (15.3.5) as

$$(L^*(y_u - y_d) + u,\ v-u)_\Sigma \geqslant 0, \qquad v \in U_{ad}. \tag{15.3.6}$$

L^* is expressed by introducing the *costate* p as follows:

for $y \in L^2(Q)$ we define p as the generalized solution of

$$\begin{aligned} -\frac{d}{dt}p - \Delta p &= y \text{ in } Q,\\ \frac{\partial p}{\partial \nu} &= 0 \text{ on } \Sigma,\\ p(T) &= 0. \end{aligned} \tag{15.3.7}$$

Then $L^*y = \gamma_0 p$.

Observe that Lv is given by

$$\begin{aligned} \frac{d}{dt}(Lv) - \Delta(Lv) &= 0 \text{ in } Q,\\ \frac{\partial}{\partial \nu}(Lv) &= v \text{ on } \Sigma,\\ (Lv)(0) &= 0, \end{aligned}$$

i.e.

$$\frac{d}{dt}((Lv)(t),\psi)_\Omega + a((Lv)(t),\psi) = (v(t),\gamma_0\psi)_\Gamma \tag{15.3.8}$$

for each $\psi \in H^1(\Omega)$, and $(Lv)(0) = 0$.

The costate p satisfies

$$-\frac{d}{dt}(p(t),\psi)_\Omega + a(p(t),\psi) = (y(t),\psi)_\Omega \tag{15.3.9}$$

for $\psi \in H^1(\Omega)$, and $p(T) = 0$.

Now,

$$
\begin{aligned}
0 &= (p(T),(Lv)(T))_\Omega - (p(0),(Lv)(0))_\Omega \\
&= \int_0^T \frac{d}{dt}(p(t),(Lv)(t))_\Omega \, dt \\
&= \int_0^T (\frac{dp(t)}{dt}, (Lv)(t))_\Omega \, dt + \int_0^T (p(t), \frac{d}{dt}(Lv)(t))_\Omega \, dt \\
&= \int_0^T a(p(t),(Lv)(t))dt - \int_0^T (y(t),(Lv)(t))_\Omega dt \qquad (15.3.10) \\
&\quad - \int_0^T a((Lv)(t), p(t))dt + \int_0^T (v(t), \gamma_0 p(t))_\Gamma \, dt,
\end{aligned}
$$

by (15.3.8) and (15.3.9). Since $a(\varphi,\psi) = a(\psi,\varphi)$, (15.3.10) yields

$$(y, Lv)_\Omega = (v, \gamma_0 p)_\Sigma$$

for $y \in L^2(Q)$, $v \in L^2(\Sigma)$. Thus $L^*v = \gamma_0 p$ as claimed.

We therefore obtain from (15.3.6) that the optimal control $u \in U_{ad}$ is characterized by

$$(\gamma_0 p_u + u, v - u)_\Sigma \geq 0 \quad \text{for all } v \in U_{ad},$$

where p_u is the generalized solution of

$$
\begin{aligned}
&-\frac{d}{dt} p_u - \Delta p_u = y_u - y_d \text{ in } Q, \\
&\frac{\partial p_u}{\partial \nu} = 0 \text{ on } \Sigma, \\
&p_u(T) = 0.
\end{aligned}
$$
■

In the case of no constraints on the controls ($U_{ad} = L^2(\Sigma)$), (15.3.1) yields

$$u = -\gamma_0 p_u, \tag{15.3.11}$$

and we can eliminate u from (15.3.2) and (15.3.3) and obtain for the optimal state y and costate p the 'coupled' system of equations

$$\begin{aligned} &\frac{dy}{dt} - \Delta y = 0 \text{ in } Q \\ &-\frac{dp}{dt} - \Delta p = y - y_d \text{ in } Q, \\ &\frac{\partial y}{\partial \nu} = -p, \quad \frac{\partial p}{\partial \nu} = 0 \text{ on } \Sigma, \\ &y(0) = \overset{0}{y}, \quad p(T) = 0. \end{aligned} \tag{15.3.12}$$

If $\{y,p\}$ solves (15.3.12), the optimal control u is given by

$$u = -\gamma_0 p. \tag{15.3.13}$$

As in the finite-dimensional case, we should like to 'decouple' the system (15.3.12), express p as

$$p(t) = P(t)y(t) + r(t), \quad t \in (0,T), \tag{15.3.14}$$

and thus obtain from (15.3.13) the optimal control in feedback form.

15.4 DECOUPLING

In this section we briefly outline Lions's indirect study of the Riccati-type equation satisfied by P(t) of (15.3.14).

The first step is to observe that we can describe P(s) and r(s) by considering the optimal control problem on (s,T), $0 \leqslant s < T$, with arbitrary initial time s and corresponding initial state $h \in L^2(\Omega)$. Specifically, one considers the counterpart of (P) on (s,T) with $y_d = 0$ and initial state $h \in L^2(\Omega)$.

For the corresponding optimal state β and costate γ we have the following system:

$$\frac{d\beta}{dt} - \Delta\beta = 0 \text{ on } \Omega \times (s,T),$$

$$-\frac{d\gamma}{dt} - \Delta\gamma = \beta \text{ on } \Omega \times (s,T), \tag{15.4.1}$$

$$\frac{\partial\beta}{\partial\nu} = -\gamma \text{ , } \frac{\partial\gamma}{\partial\nu} = 0 \text{ on } \Gamma \times (s,T),$$

$$\beta(s) = h, \quad \gamma(T) = 0.$$

Then $P(s)h = \gamma(s)$. For details see [6, Chapter III]. The function $r(s)$ is obtained by considering the same type of problem on (s,T) with initial state $h=0$. Thus, $r(s) = \xi(s)$, where ξ is the optimal costate obtained from

$$\frac{d\eta}{dt} - \Delta\eta = 0 \text{ in } \Omega \times (s,T),$$

$$-\frac{d\xi}{dt} - \Delta\xi = \eta - y_d \text{ in } \Omega \times (s,T), \tag{15.4.2}$$

$$\frac{\partial\eta}{\partial\nu} = -\xi, \quad \frac{\partial\xi}{\partial\nu} = 0 \text{ on } \Gamma \times (s,T),$$

$$\eta(s) = 0, \quad \xi(T) = 0.$$

The characterization of $P(s)$ and $r(s)$ in terms of the systems (15.4.1) and (15.4.2) enabled Lions [6] to establish certain fundamental properties of $P(s)$ and $r(s)$, which we state without proof.

THEOREM 15.4.1 *The operator* $P(s) \in \mathcal{L}(L^2(\Omega), L^2(\Omega))$ *is positive definite and self-adjoint. Furthermore* $P(s) \in \mathcal{L}(L^2(\Omega), H^1(\Omega))$, *and the function* $s \longmapsto P(s)h$ $([0,T] \to H^1(\Omega))$ *is continuously differentiable for all* $h \in L^2(\Omega)$. *We have* $r \in L^2(0,T;H^1(\Omega))$.

Using this result, the following theorem is obtained.

THEOREM 15.4.2 The operator P(t) *satisfies*

$$-\frac{d}{dt}(P(t)\varphi,\psi)_\Omega + a(\dot{\varphi},P(t)\psi) + a(P(t)\varphi,\psi)$$

$$+ (P(t)\varphi,\ P(t)\psi)_\Gamma = (\varphi,\psi)_\Omega \tag{15.4.3}$$

for all $\varphi,\psi \in H^1(\Omega)$, and $P(T) = 0$.

The function r(t) *satisfies*

$$-\frac{d}{dt}(r(t),\psi)_\Omega + a(r(t),\varphi) + (r(t),\ P(t)\psi)_\Omega$$

$$= (-y_d,\psi)_\Omega, \tag{15.4.4}$$

for all $\psi \in H^1(\Omega)$, and $r(T) = 0$.

The direct study of the Riccati-type equation satisfied by P(t) is largely an open field of research (see, however, Russell's paper [8] concerning the control of a hyperbolic equation in one space variable). Nevertheless, an indirect study along the lines of Lions [6] may give useful information about the solution (see, for example, the paper by Vinter and Johnson [9]). Such information can clarify questions on the convergence of approximate methods.

15.5 SOME COMMENTS

The example treated in previous sections illustrates Lions's approach to the theory of the optimal control of systems governed by partial differential equations. We may call it the variational approach, since the equations are treated in 'variational' form, i.e. solutions are generalized solutions, as in our example. Another general framework for the study of partial differential equations is the semigroup-theoretic approach. Optimal control may also be studied within the latter framework [1],[3],[4].

For questions concerning controllability, the survey article by Russell [7] may be consulted. As is to be expected, there is no complete theory like that for finite-dimensional systems, and there are many open pro= blems. For some results on numerical approximation see [2],[5].

15.6 REFERENCES

[1] A.V. Balakrishnan, *Applied Functional Analysis*. Springer-Verlag, Berlin, 1976.

[2] A.Bensoussan, A. Bossavit, J.C. Nedelec, *Approximation des Problemes de Control Optimal*. Cahier no.2, Institut de Recherche d'Infor= matique et d'Automatique, 1970.

[3] R.F. Curtain, A.J. Pritchard, *Infinite Dimensional Linear Systems Theory*, Springer-Verlag, 1978.

[4] R.F. Curtain, A.J. Pritchard, An abstract theory for unbounded control action for distributed parameter systems, *SIAM J. Control and Optimization*, Vol 15, 1977, pp 566-611.

[5] W. Hackbusch, On the fast solving of parabolic boundary control problems, *SIAM J Control and Optimization*, Vol 17, 1979, pp 231-244.

[6] J.L. Lions, *Optimal Control of Systems Governed by Partial Differen= tial Equations*, Springer-Verlag, Berlin, 1971.

[7] D.L. Russell, Controllability and stabilizability theory for linear partial differential equations : recent progress and open questions, *SIAM Review*, Vol 20, 1978, pp 639-739.

[8] D.L. Russell, Quadratic performance criteria in boundary control of linear symmetric hyperbolic systems, *SIAM J. Control and Optimization*, Vol 11, 1973, pp 475-509.

[9] R.B. Vinter, T.L. Johnson, Optimal control of nonsymmetric hyperbolic systems in n variables on the half-space. *SIAM J. Control and Optimization*, Vol 15, 1977, pp 129-143.